The Jacknife, the Bootstrap and Other Resampling Plans

CBMS-NSF REGIONAL CONFERENCE SERIES IN APPLIED MATHEMATICS

A series of lectures on topics of current research interest in applied mathematics under the direction of the Conference Board of the Mathematical Sciences, supported by the National Science Foundation and published by SIAM.

GARRETT BIRKHOFF, *The Numerical Solution of Elliptic Equations*
D. V. LINDLEY, *Bayesian Statistics, A Review*
R. S. VARGA, *Functional Analysis and Approximation Theory in Numerical Analysis*
R. R. BAHADUR, *Some Limit Theorems in Statistics*
PATRICK BILLINGSLEY, *Weak Convergence of Measures: Applications in Probability*
J. L. LIONS, *Some Aspects of the Optimal Control of Distributed Parameter Systems*
ROGER PENROSE, *Techniques of Differential Topology in Relativity*
HERMAN CHERNOFF, *Sequential Analysis and Optimal Design*
J. DURBIN, *Distribution Theory for Tests Based on the Sample Distribution Function*
SOL I. RUBINOW, *Mathematical Problems in the Biological Sciences*
P. D. LAX, *Hyperbolic Systems of Conservation Laws and the Mathematical Theory of Shock Waves*
I. J. SCHOENBERG, *Cardinal Spline Interpolation*
IVAN SINGER, *The Theory of Best Approximation and Functional Analysis*
WERNER C. RHEINBOLDT, *Methods of Solving Systems of Nonlinear Equations*
HANS F. WEINBERGER, *Variational Methods for Eigenvalue Approximation*
R. TYRRELL ROCKAFELLAR, *Conjugate Duality and Optimization*
SIR JAMES LIGHTHILL, *Mathematical Biofluiddynamics*
GERARD SALTON, *Theory of Indexing*
CATHLEEN S. MORAWETZ, *Notes on Time Decay and Scattering for Some Hyperbolic Problems*
F. HOPPENSTEADT, *Mathematical Theories of Populations: Demographics, Genetics and Epidemics*
RICHARD ASKEY, *Orthogonal Polynomials and Special Functions*
L. E. PAYNE, *Improperly Posed Problems in Partial Differential Equations*
S. ROSEN, *Lectures on the Measurement and Evaluation of the Performance of Computing Systems*
HERBERT B. KELLER, *Numerical Solution of Two Point Boundary Value Problems*
J. P. LASALLE, *The Stability of Dynamical Systems*
D. GOTTLIEB AND S. A. ORSZAG, *Numerical Analysis of Spectral Methods: Theory and Applications*
PETER J. HUBER, *Robust Statistical Procedures*
HERBERT SOLOMON, *Geometric Probability*
FRED S. ROBERTS, *Graph Theory and Its Applications to Problems of Society*
JURIS HARTMANIS, *Feasible Computations and Provable Complexity Properties*
ZOHAR MANNA, *Lectures on the Logic of Computer Programming*
ELLIS L. JOHNSON, *Integer Programming: Facets, Subadditivity, and Duality for Group and Semi-Group Problems*
SHMUEL WINOGRAD, *Arithmetic Complexity of Computations*
J. F. C. KINGMAN, *Mathematics of Genetic Diversity*
MORTON E. GURTIN, *Topics in Finite Elasticity*
THOMAS G. KURTZ, *Approximation of Population Processes*
JERROLD E. MARSDEN, *Lectures on Geometric Methods in Mathematical Physics*
BRADLEY EFRON, *The Jackknife, the Bootstrap, and Other Resampling Plans*
M. WOODROOFE, *Nonlinear Renewal Theory in Sequential Analysis*
D. H. SATTINGER, *Branching in the Presence of Symmetry*
R. TEMAM, *Navier–Stokes Equations and Nonlinear Functional Analysis*

Miklós Csörgo, *Quantile Processes with Statistical Applications*
J. D. Buckmaster and G. S. S. Ludford, *Lectures on Mathematical Combustion*
R. E. Tarjan, *Data Structures and Network Algorithms*
Paul Waltman, *Competition Models in Population Biology*
S. R. S. Varadhan, *Large Deviations and Applications*
Kiyosi Itô, *Foundations of Stochastic Differential Equations in Infinite Dimensional Spaces*
Alan C. Newell, *Solitons in Mathematics and Physics*
Pranab Kumar Sen, *Theory and Applications of Sequential Nonparametrics*
László Lovász, *An Algorithmic Theory of Numbers, Graphs and Convexity*
E. W. Cheney, *Multivariate Approximation Theory: Selected Topics*
Joel Spencer, *Ten Lectures on the Probabilistic Method*
Paul C. Fife, *Dynamics of Internal Layers and Diffusive Interfaces*
Charles K. Chui, *Multivariate Splines*
Herbert S. Wilf, *Combinatorial Algorithms: An Update*
Henry C. Tuckwell, *Stochastic Processes in the Neurosciences*
Frank H. Clarke, *Methods of Dynamic and Nonsmooth Optimization*
Robert B. Gardner, *The Method of Equivalence and Its Applications*
Grace Wahba, *Spline Models for Observational Data*
Richard S. Varga, *Scientific Computation on Mathematical Problems and Conjectures*
Ingrid Daubechies, *Ten Lectures on Wavelets*
Stephen F. McCormick, *Multilevel Projection Methods for Partial Differential Equations*
Harald Niederreiter, *Random Number Generation and Quasi-Monte Carlo Methods*
Joel Spencer, *Ten Lectures on the Probabilistic Method, Second Edition*
Charles A. Micchelli, *Mathematical Aspects of Geometric Modeling*
Roger Temam, *Navier–Stokes Equations and Nonlinear Functional Analysis, Second Edition*
Glenn Shafer, *Probabilistic Expert Systems*
Peter J. Huber, *Robust Statistical Procedures, Second Edition*
J. Michael Steele, *Probability Theory and Combinatorial Optimization*
Werner C. Rheinboldt, *Methods for Solving Systems of Nonlinear Equations, Second Edition*
J. M. Cushing, *An Introduction to Structured Population Dynamics*
Tai-Ping Liu, *Hyperbolic and Viscous Conservation Laws*
Michael Renardy, *Mathematical Analysis of Viscoelastic Flows*
Gérard Cornuéjols, *Combinatorial Optimization: Packing and Covering*
Irena Lasiecka, *Mathematical Control Theory of Coupled PDEs*
J. K. Shaw, *Mathematical Principles of Optical Fiber Communications*
Zhangxin Chen, *Reservoir Simulation: Mathematical Techniques in Oil Recovery*
Athanassios S. Fokas, *A Unified Approach to Boundary Value Problems*
Margaret Cheney and Brett Borden, *Fundamentals of Radar Imaging*

BRADLEY EFRON
Department of Statistics
Stanford University

The Jacknife, the Bootstrap and Other Resampling Plans

SOCIETY FOR INDUSTRIAL AND APPLIED MATHEMATICS
PHILADELPHIA

10 9 8 7

Library of Congress Catalog Card Number: 81-84708

ISBN 978-0-898711-79-0

Contents

Preface

These notes record ten lectures given at Bowling Green State University in June, 1980. The occasion was an NSF-sponsored regional conference on resampling methods, admirably organized by Professor Arjun K. Gupta. Professor Gupta and the administration of the Bowling Green Mathematics Department provided a stimulating and comfortable atmosphere for the conference, which included several other talks on jackknife-bootstrap related topics.

The lectures as they appear here have benefited from the comments of many colleagues, including several of the conference participants. I am particularly grateful to Peter Bickel, Persi Diaconis, David Hinkley, Richard Olshen and Sandy Zabell.

BRADLEY EFRON
Stanford, March 1981

CHAPTER 1

Introduction

Our goal is to understand a collection of ideas concerning the nonparametric estimation of bias, variance and more general measures of error. Historically the subject begins with the Quenouille–Tukey jackknife, which is where we will begin also. In fact it would be more logical to begin with the bootstrap, which most clearly exposes the simple idea underlying all of these methods. (And in fact underlies many common parametric methods as well, such as Fisher's information theory for assigning a standard error to a maximum likelihood estimate.) *Good* simple ideas, of which the jackknife is a prime example, are our most precious intellectual commodity, so there is no need to apologize for the easy mathematical level. The statistical ideas run deep, sometimes over our head at the current level of understanding. Chapter 10, on nonparametric confidence intervals, is particularly speculative in nature.

Some material has been deliberately omitted for these notes. This includes most of the detailed work on the jackknife, especially the asymptotic theory. Miller (1974a) gives an excellent review of the subject.

What are the jackknife and the bootstrap? It is easiest to give a quick answer in terms of a problem where neither is necessary, that of estimating the standard deviation of a sample average. The data set consists of an independent and identically distributed (i.i.d.) sample of size n from an unknown probability distribution F on the real line,

$$X_1, X_2, \cdots, X_n \overset{\text{iid}}{\sim} F. \tag{1.1}$$

Having observed values $X_1 = x_1, X_2 = x_2, \cdots, X_n = x_n$, we compute the sample average $\bar{x} = \sum_{i=1}^{n} x_i/n$, for use as an estimate of the expectation of F.

An interesting fact, and a crucial one for statistical applications, is that the data set provides more than the estimate $\bar{x}$. It also gives an estimate of the accuracy of $\bar{x}$, namely,

$$\hat{\sigma} = \left[\frac{1}{n(n-1)} \sum_{i=1}^{n} (x_i - \bar{x})^2\right]^{1/2}; \tag{1.2}$$

$\hat{\sigma}$ is the estimated standard deviation of $\bar{X}$, the root mean square error of estimation.

The trouble with formula (1.2) is that it doesn't, in any obvious way, extend to estimators other than $\bar{X}$, for example the sample median. The jackknife and the bootstrap are two ways of making this extension. Let

$$\bar{x}_{(i)}=\frac{n\bar{x}-x_i}{n-1}=\frac{1}{n-1}\sum_{j\neq i}x_j, \tag{1.3}$$

the sample average of the data set deleting the ith point. Also, let $\bar{x}_{(\cdot)}=\sum_{i=1}^{n}x_{(i)}/n$, the average of these deleted averages. Actually $\bar{x}_{(\cdot)}=\bar{x}$, but we need the dot notation below. The jackknife estimate of standard deviation is

$$\hat{\sigma}_{\text{JACK}}=\left(\frac{n-1}{n}\sum_{i=1}^{n}(\bar{x}_{(i)}-\bar{x}_{(\cdot)})^2\right)^{1/2}. \tag{1.4}$$

The reader can verify that this is the same as (1.3). The advantage of (1.4) is that it can be generalized to any estimator $\hat{\theta}=\hat{\theta}(X_1, X_2, \cdots, X_n)$. The only change is to substitute $\hat{\theta}_{(i)}=\hat{\theta}(X_1, X_2, \cdots, X_{i-1}, X_{i+1}, \cdots, X_n)$ for $\bar{x}_{(i)}$ and $\hat{\theta}_{(\cdot)}$ for $\bar{x}_{(\cdot)}$.

The bootstrap generalizes (1.3) in an apparently different way. Let $\hat{F}$ be the empirical probability distribution of the data, putting probability mass $1/n$ on each x_i, and let $X_1^*, X_2^*, \cdots, X_n^*$ be an i.i.d. sample from $\hat{F}$,

$$X_1^*, X_2^*, \cdots, X_n^* \overset{\text{iid}}{\sim} \hat{F}. \tag{1.5}$$

In other words, the X_i^* are a random sample drawn *with replacement* from the observed values $x_1, x_2, \cdots, x_n$. Then $\bar{X}^*=\sum X_i^*/n$ has variance

$$\text{Var}_* \bar{X}^*=\frac{1}{n^2}\sum_{i=1}^{n}(x_i-\bar{x})^2, \tag{1.6}$$

Var_* indicating variance under sampling scheme (1.5). The bootstrap estimate of standard deviation for an estimator $\hat{\theta}(X_1, X_2, \cdots, X_n)$ is

$$\hat{\sigma}_{\text{BOOT}}=[\text{Var}_* \hat{\theta}(X_1^*, X_2^*, \cdots, X_n^*)]^{1/2}. \tag{1.7}$$

Comparing (1.7) with (1.2), we see that $\hat{\sigma}_{\text{BOOT}}=[(n-1)/n]^{1/2}\hat{\sigma}$ for $\hat{\theta}=\bar{X}$. We could make $\hat{\sigma}_{\text{BOOT}}$ exactly equal to $\hat{\sigma}$, for $\hat{\theta}=\bar{X}$, simply by adjusting definition (1.7) with the factor $[n/(n-1)]^{1/2}$, but there turns out to be no advantage in doing so. A simple algorithm described in Chapter 4 allows the statistician to compute $\hat{\sigma}_{\text{BOOT}}$ no matter how complicated $\hat{\theta}$ may be.

Many other generalizations of (1.3) have been put forth. All such methods turn out to be closely related in theory, but not necessarily in their numerical consequences for a specific data set. We shall be investigating the theoretical and practical aspects of this collection of methods. Other measures of statistical error besides standard deviation are also considered: bias, prediction error and confidence intervals.

From a traditional point of view, all of the methods discussed here are prodigious computational spendthrifts. We blithely ask the reader to consider techniques which require the usual statistical calculations to be multiplied a thousand times over. None of this would have been feasible twenty-five years ago, before the era of cheap and fast computation. An important theme of what follows is the substitution of computational power for theoretical analysis. The payoff, of course, is freedom from the constraints of traditional parametric theory, with its overreliance on a small set of standard models for which theoretical solutions are available. In the long run, understanding the limitations of the nonparametric approach should make clearer the virtues of parametric theory, and perhaps suggest useful compromises. Some hints of this appear in Chapter 10, but so far these are only hints and not a well developed point of view.

CHAPTER 2

The Jackknife Estimate of Bias

Quenouille (1949) invented a nonparametric estimate of bias, subsequently named the jackknife, which is the subject of this chapter. Suppose that we sample independent and identically distributed random quantities $X_1, X_2, X_3, \cdots, X_n \overset{\text{iid}}{\sim} F$, where F is an unknown probability distribution on some space $\mathscr{X}$. Often $\mathscr{X}$ will be the real line, but all of the methods discussed here allow $\mathscr{X}$ to be completely arbitrary. Having observed $X_1 = x_1, X_2 = x_2, \cdots, X_n = x_n$, we compute some statistic of interest, say

$$\hat{\theta} = \hat{\theta}(x_1, x_2, \cdots \ x_n).$$

We are interested in the bias of $\hat{\theta}$ for estimating a true quantity θ.

For now we concentrate on *functional statistics*: $\theta(F)$ is some real-valued parameter of interest, such as an expectation, a quantile, a correlation, etc., which we estimate by the statistic

$$\hat{\theta} = \theta(\hat{F}), \tag{2.1}$$

where $\hat{F}$ is the *empirical probability distribution*,

$$\hat{F}: \text{mass}\,\frac{1}{n} \quad \text{at } x_1, x_2, \cdots, x_n. \tag{2.2}$$

Form (2.1) guarantees that $\hat{\theta}(x_1, x_2, \cdots, x_n)$ is invariant under permutations of the arguments, which we use below, and more importantly that the concept of bias is well defined,

$$\text{Bias} \equiv E_F \theta(\hat{F}) - \theta(F). \tag{2.3}$$

Here "E_F" indicates expectation under $X_1, X_2, \cdots, X_n \overset{\text{iid}}{\sim} F$. Three familiar examples of functional statistics follow.

Example 2.1. *The expectation.* $\mathscr{X} = \mathscr{R}^1$ the real line; $\theta(F) = \int_{\mathscr{X}} x\, dF = E_F X$; $\hat{\theta} = \int_{\mathscr{X}} x\, d\hat{F} = (1/n) \sum x_i = \bar{x}$, the average, or sample expectation.

Example 2.2. *The correlation.* $\mathscr{X} = \mathscr{R}^2$ the plane; $\theta(F) =$ Pearson's product-moment correlation coefficient (Cramér (1946, p. 265)); $\hat{\theta} = \theta(\hat{F})$, the sample correlation coefficient.

Example 2.3. *Ratio estimation.* $\mathscr{X} = \mathscr{R}^{2+}$ the positive quadrant of the plane; with $X = (Y, Z)$, then $\theta(F) = E_F(Z)/E_F(Y)$, the ratio of expectations for the two coordinates; $\hat{\theta} = \theta(\hat{F}) = \bar{z}/\bar{y}$, the ratio of corresponding averages.

2.1. Quenouille's bias estimate. Quenouille's method is based on sequentially deleting points x_i, and recomputing $\hat{\theta}$. Removing point x_i from the data set gives a different empirical probability distribution,

$$\hat{F}_{(i)}: \text{mass } \frac{1}{n-1} \quad \text{at } x_1, x_2, \cdots, x_{i-1}, x_{i+1}, \cdots, x_n, \tag{2.4}$$

and a corresponding recomputated value of the statistic,

$$\hat{\theta}_{(i)} = \theta(\hat{F}_{(i)}) = \hat{\theta}(x_1, x_2, \cdots, x_{i-1}, x_{i+1}, \cdots, x_n). \tag{2.5}$$

Let

$$\hat{\theta}_{(\cdot)} = \frac{1}{n} \sum_{i=1}^{n} \hat{\theta}_{(i)}. \tag{2.6}$$

Quenouille's estimate of bias is

$$\widehat{\text{BIAS}} \equiv (n-1)(\hat{\theta}_{(\cdot)} - \hat{\theta}), \tag{2.7}$$

leading to the bias-corrected "jackknifed estimate" of θ

$$\tilde{\theta} \equiv \hat{\theta} - \widehat{\text{BIAS}} = n\hat{\theta} - (n-1)\hat{\theta}_{(\cdot)}. \tag{2.8}$$

The usual rationale for $\widehat{\text{BIAS}}$ and $\tilde{\theta}$ goes as follows. If E_n denotes the expectation for sample size n, $E_n \equiv E_F\hat{\theta}(X_1, X_2, \cdots, X_n)$, then for many common statistics, including most maximum likelihood estimates,

$$E_n = \theta + \frac{a_1(F)}{n} + \frac{a_2(F)}{n^2} + \cdots, \tag{2.9}$$

where the functions $a_1(F), a_2(F), \cdots$ do not depend upon n; see Schucany, Gray and Owen (1971). Notice that

$$E_F\hat{\theta}_{(\cdot)} = E_{n-1} = \theta + \frac{a_1(F)}{n-1} + \frac{a_2(F)}{(n-1)^2} + \cdots,$$

and so

$$E_F\tilde{\theta} = nE_n - (n-1)E_{n-1} = \theta - \frac{a_2(F)}{n(n-1)} + a_3(F)\left(\frac{1}{n^2} + \frac{1}{(n-1)^2}\right) + \cdots. \tag{2.10}$$

We see that $\tilde{\theta}$ is biased $O(1/n^2)$ compared to $O(1/n)$ for the original estimator.

Example 2.1 *continued*. For $\hat{\theta} = \bar{x}$ we calculate $\hat{\theta}_{(\cdot)} = \bar{x} = \hat{\theta}$ and $\widehat{\text{BIAS}} = 0$. Of course $\hat{\theta}$ is unbiased for $\theta = E_FX$ in this case, so $\widehat{\text{BIAS}} = 0$ is the correct bias estimate.

Example 2.4. *The variance*. $\mathscr{X} = \mathscr{R}^1$; $\theta(F) = \int_{\mathscr{X}} (x - E_FX)^2 \, dF$; $\hat{\theta} = \sum_{i=1}^{n} (x_i - \bar{x})^2/n$. A simple calculation shows that

$$\widehat{\text{BIAS}} = -\frac{1}{n(n-1)} \sum_{i=1}^{n} (x_i - \bar{x})^2,$$

yielding

$$\tilde{\theta} = \frac{1}{n-1} \sum_{i=1}^{n} (x_i - \bar{x})^2,$$

the usual unbiased estimate of θ. In this case the expansion (2.9) is

$$E_n = \theta - \frac{\theta}{n},$$

so $a_1(F) = -\theta$, $a_j(F) = 0$ for $j > 1$. The motivating formula (2.10) shows why $\tilde{\theta}$ is exactly unbiased in this situation.

2.2. The grouped jackknife. Suppose $n = gh$ for integers g and h. We can remove observations in blocks of size h, e.g., first remove $x_1, x_2, \cdots, x_h$, second remove $x_{h+1}, x_{h+2}, \cdots, x_{2h}$, etc. Now define $\hat{\theta}_{(i)}$ as the statistic recomputed with the ith block removed, and $\hat{\theta}_{(\cdot)} = (1/g)\sum \hat{\theta}_{(i)}$. Then

$$\tilde{\theta} = g\hat{\theta} - (g-1)\hat{\theta}_{(\cdot)}$$

also removes the first order of the bias term, as in (2.10). Quenouille (1949) considered the "half-sample" case $g = 2$.

If computationally feasible, it is preferable to define $\hat{\theta}_{(\cdot)} = \sum_{\mathbf{i}} \hat{\theta}_{(\mathbf{i})} / \binom{n}{h}$, where $\mathbf{i}$ indicates a subset of size h removed from $\{1, 2, \cdots, n\}$, and $\sum_{\mathbf{i}}$ is the sum over all such subsets. Then $\tilde{\theta} = g\hat{\theta} - (g-1)\hat{\theta}^{(\cdot)}$ has the same expectation and smaller variance than in the blocked case above, by a sufficiency argument. We consider only the ungrouped jackknife, $h = 1$, in what follows, except for a few occasional remarks.

2.3. A picture. Figure 2.1 shows E_n graphed versus $1/n$. The notation $\theta = E_\infty$ is based on (2.9), with $n = \infty$. Assuming perfect linearity of the bias in $1/n$ implies

$$\frac{E_n - E_\infty}{E_{n-1} - E_n} = \frac{1/n}{1/(n-1) - 1/n},$$

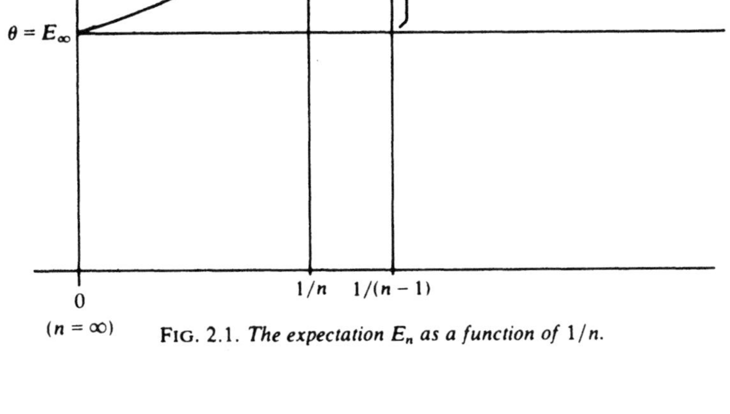

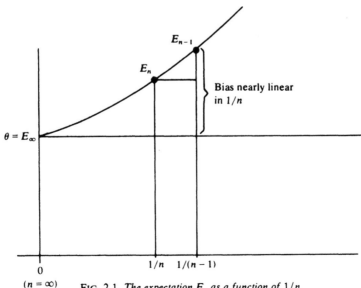

FIG. 2.1. *The expectation E_n as a function of $1/n$.*

which gives

$$\text{Bias} = E_n - E_\infty = (n-1)(E_{n-1} - E_n)$$

and

$$\theta = E_\infty = nE_n - (n-1)E_{n-1}. \tag{2.11}$$

The jackknife formulas (2.7), (2.8) simply replace E_n and E_{n-1} on the right side of (2.11) by their unbiased estimates $\hat{\theta}$ and $\hat{\theta}_{(\cdot)}$ respectively.

Removing two data points at a time and averaging the resulting recomputed values of $\hat{\theta}$ gives an unbiased estimate of E_{n-2}, say $\hat{\theta}_{(\cdot\cdot)}$. Looking at Fig. 2.1, it seems reasonable to use quadratic extrapolation to predict $\theta = E_\infty$ from $\hat{\theta}$, $\hat{\theta}_{(\cdot)}$, $\hat{\theta}_{(\cdot\cdot)}$. A neat way of deriving all such higher order bias correction formulas is given in Schucany, Gray and Owen (1971).

In practice there has been little use made of higher order bias correction. Even the first order bias correction (2.8) may add more to the mean square error in variance than it removes in $(\text{bias})^2$. Hinkley (1978) discusses this effect for the case of the correlation coefficient. However it can still be interesting to compute $\widehat{\text{BIAS}}$, even if the bias correction is not made, especially in conjunction with $\widehat{\text{VAR}}$, an estimate of variance such as that discussed in Chapter 3. Very often it turns out that $\widehat{\text{BIAS}}/\sqrt{\widehat{\text{VAR}}}$ is small, say $<\frac{1}{4}$, in which case bias is probably not a serious issue.

2.4. Aitken acceleration. The extrapolation method underlying the jackknife has a long history in numerical analysis, two examples being Aitken acceleration and Richardson extrapolation. (Here we discuss only the former.) The connection is as follows. If we let $E_n - E_\infty = \text{Bias}_n$, the bias for sample size n, simple algebra yields

$$E_\infty = \frac{E_n - (\text{ratio})E_{n-1}}{1 - (\text{ratio})},$$

where

$$\text{ratio} = \frac{\text{Bias}_n}{\text{Bias}_{n-1}}.$$

The approximation $\text{Bias}_n/\text{Bias}_{n-1} \doteq (n-1)/n$ gives (2.11), and hence the jackknife results (2.7), (2.8).

Now suppose we wish to approximate an infinite sum $S_\infty = \sum_{k=0}^{\infty} b_k$ on the basis of finite sums $S_n = \sum_{k=0}^{n} b_k$. With $B_n = \sum_{n+1}^{\infty} b_k$, the "bias" in S_n for approximating S, the same simple algebra yields

$$S_\infty = \frac{S_n - (\text{ratio})S_{n-1}}{1 - (\text{ratio})},$$

where

$$\text{ratio} = \frac{B_n}{B_{n-1}}.$$

Aitken acceleration replaces B_n/B_{n-1} with $b_n/b_{n-1} = (S_n - S_{n-1})/(S_{n-1} - S_{n-2})$, which is exactly right for a geometric series $b_k = cr^k$. The series transformation

$$\tilde{S}_n = \frac{S_n - \dfrac{S_n - S_{n-1}}{S_{n-1} - S_{n-2}} S_{n-1}}{1 - \dfrac{S_n - S_{n-1}}{S_{n-1} - S_{n-2}}} \tag{2.12}$$

can be applied repeatedly to speed up convergence. As an example, borrowed from Gray, Watkins and Adams (1972), consider the series $\pi = 4 - 4/3 + 4/5 - 4/7 + \cdots$. Taking only seven terms of the original series, and applying (2.12) three times, gives 3.14160, with deviation less than .00001 from $S_\infty = 3.141593 \cdots$. See Table 2.1.

TABLE 2.1

n	$\tilde{S}_n$	$\tilde{S}_n^{(1)}$	$\tilde{S}_n^{(2)}$	$\tilde{S}_n^{(3)}$
0	4.00000			
1	2.66667			
2	3.46667	3.16667		
3	2.89524	3.13334		
4	3.33968	3.14524	3.14211	
5	2.97605	3.13968	3.14145	
6	3.28374	3.14271	3.14164	3.14160

Iterating (2.12) three times amounts to using a cubic extrapolation formula in Fig. 2.1. This is more reasonable in a numerical analysis setting than in the noisy world of statistical estimation.

2.5. The law school data. Table 2.2 gives the average LSAT and GPA scores for the 1973 entering classes of 15 American law schools. (LSAT is a national test for prospective lawyers, GPA the undergraduate grade point average; see Efron (1979b) for details.) The data are plotted in Fig. 2.2. Consider Pearson's correlation coefficient, as in Example 2.2. Denoting the statistic by $\hat{\rho}$ rather than by $\hat{\theta}$, gives the value of the sample correlation coefficient for the 15 schools as $\hat{\rho} = .776$. The quantities $\hat{\rho}_{(i)} - \hat{\rho}$, also given in Table 2.2, yield $\widehat{\text{BIAS}} = 14(\hat{\rho}_{(\cdot)} - \hat{\rho}) = -.007$. As a point of comparison, the normal theory estimate of Bias is $-.011$, obtained by substituting $\hat{\rho}$ in formula 10.1 from Johnson and Kotz (1970, p. 225). We will return to this example several times in succeeding chapters.

TABLE 2.2.
Average LSAT and GPA for the 1975 entering classes of 15 American law schools. Values of $\hat{\rho}_{(i)}-\hat{\rho}$ are used to compute $\widehat{\text{BIAS}}$ for the correlation coefficient.

School	1	2	3	4	5	6	7	8
LSAT	576	635	558	578	666	580	555	661
GPA	3.39	3.30	2.81	3.03	3.44	3.07	3.00	3.43
$\hat{\rho}_{(i)}-\hat{\rho}$	.116	−.013	−0.021	−.000	−.045	.004	.008	−.040

School	9	10	11	12	13	14	15
LSAT	651	605	653	575	545	572	594
GPA	3.36	3.13	3.12	2.74	2.76	2.88	2.96
$\hat{\rho}_{(i)}-\hat{\rho}$	−.025	−.000	.042	.009	−.036	−.009	.003

2.6. What does $\widehat{\text{BIAS}}$ really estimate? The motivation for Quenouille's estimate $\widehat{\text{BIAS}} = (n-1)(\hat{\theta}_{(\cdot)} - \hat{\theta})$ based on (2.9) collapses under closer scrutiny. Notice that (2.9) can always be rewritten as

$$E_n = \theta + \frac{A_1(F)}{n+1} + \frac{A_2(F)}{(n+1)^2} + \cdots,$$

where $A_1(F) = a_1(F)$, $A_2(F) = a_1(F) + a_2(F)$, $A_3(F) = a_1(F) + 2a_2(F) + a_3(F)$, $\cdots$. The same reasoning that gave (2.10) now gives

$$E_F\{(n+1)\hat{\theta} - n\hat{\theta}_{(\cdot)}\} = \theta - \frac{A_2(F)}{n(n+1)} + A_3(F)\left(\frac{1}{(n+1)^3} - \frac{1}{n^3}\right)\cdots.$$

This would suggest that $\widehat{\text{BIAS}} = n(\hat{\theta}_{(\cdot)} - \hat{\theta})$ is the correct formula for removing the $1/n$ bias term, not Quenouille's formula (2.7).

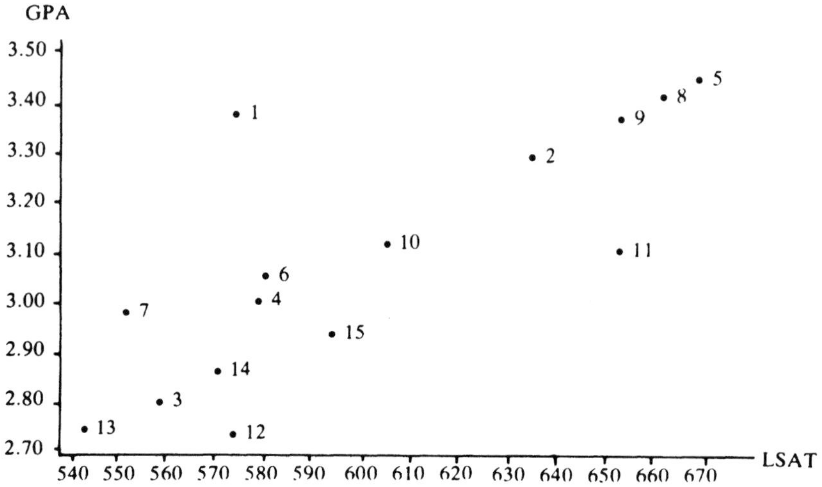

FIG. 2.2. *A plot of the law school data given in Table 2.2.*

The real justification for Quenouille's formula is contained in Example 2.4. The statistic $\hat{\theta}=\sum(x_i-\bar{x})^2/n$ is an example of a *quadratic functional*: $\hat{\theta}$ is of the functional form (2.1), $\hat{\theta}=\theta(\hat{F})$, and $\hat{\theta}$ can be expressed as

$$\hat{\theta}=\mu^{(n)}+\frac{1}{n}\sum_{i=1}^{n}\alpha^{(n)}(x_i)+\frac{1}{n^2}\sum_{1\leqq i_1<i_2\leqq n}\beta(x_{i_1},x_{i_2}), \tag{2.13}$$

i.e., in a form which involves the x_i one and two at a time, but in no higher order interactions. Quadratic functionals, which are closely related to Hoeffding's (1948) U-statistics, are discussed in Chapter 4. The proof of the following theorem is given there.

THEOREM 2.1. *For a quadratic functional,* $\widehat{\text{BIAS}}=(n-1)(\hat{\theta}_{(\cdot)}-\hat{\theta})$ *is unbiased for estimating the true bias* $E_F\theta(\hat{F})-\theta(F)$.

It is easy to think of functional statistics $\hat{\theta}=\theta(\hat{F})$ for which $\widehat{\text{BIAS}}$ is useless or worse than useless. For example, let $\mathscr{X}=\mathscr{R}^1$, and $\theta(F)=0$ if F has no discrete probability atoms, $\theta(F)=1$ if it does. If F has no atoms, so that $\theta=0$, then $\hat{\theta}=\theta(\hat{F})=1$ has true bias 1, while $\widehat{\text{BIAS}}=0$ for any sample size n. This points out that the concept (2.1) of a functional statistic is itself useless without some notion of continuity in the argument $\hat{F}$; see Huber (1977, Chapt. II). Its only purpose in this chapter was to give an unambiguous meaning to the concept of bias. We will see what $\widehat{\text{BIAS}}$ actually estimates, whether or not the statistic is functional, at the end of Chapter 6. Roughly speaking, $\widehat{\text{BIAS}}$ is the true bias of $\hat{\theta}$ if F were actually equal to the observed $\hat{F}$. That is, $\widehat{\text{BIAS}}$ itself is (approximately) a functional statistic, the function being the bias of $\hat{\theta}$. All of this will be made clear in Chapter 5 when we discuss the bootstrap.

CHAPTER 3

The Jackknife Estimate of Variance

Tukey (1958) suggested how the recomputed statistics $\hat{\theta}_{(i)}$ could also provide a nonparameteric estimate of variance. Let

$$\text{Var} = E_F[\hat{\theta}(X_1, X_2, \cdots, X_n) - E_F\hat{\theta}]^2, \tag{3.1}$$

where as before E_F indicates expectation with $X_1, X_2, \cdots, X_n \overset{\text{iid}}{\sim} F$, F an unknown probability distribution on some space $\mathscr{X}$. (In general, "E_F" means that all random variables involved in the expectation are independently distributed according to F.) Tukey's formula for estimating Var is[1]

$$\widehat{\text{VAR}} = \frac{n-1}{n} \sum_{i=1}^{n} [\hat{\theta}_{(i)} - \hat{\theta}_{(\cdot)}]^2, \tag{3.2}$$

$\hat{\theta}_{(\cdot)} = \sum \hat{\theta}_{(i)}/n$. We will often be more interested in standard deviations than variances, since the standard deviation relates directly to accuracy statements about $\hat{\theta}$, in which case we will use the notation

$$\text{Sd} = \sqrt{\text{Var}}, \qquad \text{SD} = \sqrt{\widehat{\text{VAR}}}. \tag{3.3}$$

Considerable effort has gone into verifying, and in some cases disverifying, the usefulness of $\widehat{\text{VAR}}$ as an estimate of Var; see Miller (1974b). This chapter presents several examples typifying $\widehat{\text{VAR}}$'s successes and failures. The theoretical basis of (3.2) is discussed in Chapters 4, 5 and 6.

3.1. The expectation. As in Example 2.1, $\hat{\theta} = \bar{x} = \sum x_i/n$. Then

$$\hat{\theta}_{(i)} = \frac{n\hat{\theta} - x_i}{n-1}, \quad \hat{\theta}_{(\cdot)} = \hat{\theta}, \quad \hat{\theta}_{(i)} - \hat{\theta}_{(\cdot)} = \frac{\bar{x} - x_i}{n-1},$$

and so

$$\widehat{\text{VAR}} = \frac{1}{n(n-1)} \sum (x_i - \bar{x})^2,$$

the usual nonparametric estimate for the variance of an average $\bar{X}$. This is the motivating example behind Tukey's formula (3.2).

Pseudo-values. Tukey called

$$\tilde{\theta}_i = \hat{\theta} + (n-1)(\hat{\theta} - \hat{\theta}_{(i)})$$

[1] Some writers consider $\widehat{\text{VAR}}$ as an estimate of $\text{Var}(\tilde{\theta})$, rather than of $\text{Var}(\hat{\theta})$, but in fact it seems to be a better estimator of the latter; see Hinkley (1978). This will be our point of view.

the "ith pseudo-value". For a general statistic $\hat{\theta}$, the jackknife estimate $\tilde{\theta}$ equals $\sum \tilde{\theta}_i/n$, and $\widehat{\text{VAR}} = \sum (\tilde{\theta}_i - \tilde{\theta})^2/[n \cdot (n-1)]$. This makes the $\tilde{\theta}_i$ appear to be playing the same role as do the x_i in the case $\hat{\theta} = \bar{x}$. (Indeed, $\tilde{\theta}_i = x_i$ when $\hat{\theta} = \bar{x}$.)

Unfortunately the analogy doesn't seem to go deep enough. Attempts to extract additional information from the $\tilde{\theta}_i$ values, beyond the estimate $\widehat{\text{VAR}}$, have been disappointing. For example Tukey's original suggestion was to use

$$\tilde{\theta} \pm t_\alpha^{n-1} \widehat{\text{SD}} \tag{3.4}$$

as a $1-2\alpha$ confidence interval for θ, where t_α^{n-1} is the α upper percentile point of a t distribution with $n-1$ degrees of freedom. Verification of (3.4) as a legitimate confidence interval, as in Miller (1964), has been successful only in the asymptotic case $n \to \infty$, for which the "t" effect disappears, and where instead of (3.4) we are dealing with the comparatively crude limiting normal theory. (Small sample nonparametric confidence intervals are discussed in Chapter 10.) The pseudo-value terminology is slightly confusing, and will not be used in our discussion.

3.2. The unbiased estimate of variance. Let $\mathscr{X} = \mathscr{R}^1$ and $\hat{\theta} = \sum_{i=1}^n (x_i - \bar{x})^2/(n-1)$. Define the kth central moments of F and $\hat{F}$ to be, respectively,

$$\mu_k = E_F[X - E_F X]^k, \qquad \hat{\mu}_k = \frac{1}{n} \sum_{i=1}^n [x_i - \bar{x}]^k. \tag{3.5}$$

For this $\hat{\theta}$, (3.2) has the simple expression

$$\widehat{\text{VAR}} = \frac{n^2}{(n-1)(n-2)^2} (\hat{\mu}_4 - \hat{\mu}_2^2), \tag{3.6}$$

agreeing nicely with the true variance of $\hat{\theta}$,

$$\text{Var} = \frac{\mu_4 - \dfrac{n-3}{n-1} \mu_2^2}{n}. \tag{3.7}$$

Hint: it helps to work with $y_i = (x_i - \bar{x})^2$ in deriving (3.6).

Notice that this $\hat{\theta}$ is not a functional statistic since, for example, the doubled data set $x_1, x_1, x_2, x_2, \cdots, x_n, x_n$ has the same value of $\hat{F}$ but a different value of $\hat{\theta}$. Variance is simpler than bias in that it refers only to the actual sample size n. Bias refers to sample size n and also to sample size ∞, i.e., the true θ. The concept of a functional statistic plays no role in (3.2). We only assume that $\hat{\theta}(x_1, x_2, \cdots, x_n)$ is symmetrically defined in its n arguments. Chapter 6 shows that $\widehat{\text{VAR}}$ is based on a simpler idea than is $\widehat{\text{BIAS}}$, the difference being the use of a linear rather than quadratic extrapolation formula. In the author's experience, $\widehat{\text{VAR}}$ tends to yield more dependable estimates than does $\widehat{\text{BIAS}}$.

3.3. Trimmed means. Let $x_{(1)} \leqq x_{(2)} \leqq \cdots \leqq x_{(n)}$ be the order statistics of a sample on the real line $\mathscr{X} = \mathscr{R}^1$, and let α be the proportion of points we "trim" from each end of the sample. The α trimmed mean is the average of the remaining

$n(1-2\alpha)$ central order statistics. For instance if $n=10$, $\alpha=.1$, then $\hat{\theta}=\sum_2^9 x_{(i)}/8$. (The two endpoints of the trimming region are counted partially, in the obvious way, if $n\alpha$ is not an integer.) If $(n-1)\alpha=g$, an integer, then

$$\widehat{\text{VAR}}=\frac{1}{n(n-1)(1-2\alpha)^2}\sum_{i=1}^{n}[x_{(W_i)}-\bar{x}_W]^2, \tag{3.8}$$

where

$$W_i=\begin{cases} g+1, & i\leqq g+1,\\ i, & g+1<i<n-g,\\ n-g, & i\geqq n-g \end{cases}$$

and $\bar{x}_W=\sum x_{(W_i)}/n$; see Huber (1977, p. 27). The letter W stands for "Winsorized" since (3.8) is proportional to the variance of what is called the Winsorized sample, i.e., the sample where the end order statistics are not trimmed off but rather changed in value to $x_{(g+1)}$ or $x_{(n-g)}$. The expression for VAR is only slightly more complicated if $(n-1)\alpha$ is not an integer.

Formula (3.8) has proved reasonably dependable for $\alpha\leqq.25$; see Carroll (1979). Table 3.1 shows the results of a small Monte Carlo experiment: 200 trials, each consisting of a sample of size $n=15$, $X_1, X_2, \cdots, X_{15} \overset{\text{iid}}{\sim} F$. Two cases were investigated, the normal case $F\sim\mathcal{N}(0,1)$ and the negative exponential case $F\sim G_1$ (i.e., $f(x)=e^{-x}$ for $x>0$, $f(x)=0$ for $x\leqq 0$). The trimming proportion was $\alpha=.25$.

TABLE 3.1

Estimates of standard deviation for the 25% trimmed mean using the jackknife and the bootstrap: 200 trials of $X_1, X_2, \cdots, X_{15} \overset{\text{iid}}{\sim} F$. The averages and standard deviations of $\widehat{\text{SD}}$ for the 200 trials show a moderate advantage for the bootstrap.

	$F\sim\mathcal{N}(0,1)$			$F\sim G_1$		
	Ave	Std dev	Coeff var	Ave	Std dev	Coeff var
Jackknife	.280	.084	.30	.224	.085	.38
Bootstrap, 200 bootstrap reps per trial	.287	.071	.25	.242	.078	.32
True Sd [minimum possible Cv]	.286		[.19]	.232		[.27]

For $F\sim\mathcal{N}(0,1)$, the 200 jackknife estimates of standard deviation, each essentially the square root of (3.8) with $n=15$ and $\alpha=.25$, averaged .280 with sample standard deviation .084. Typically $\widehat{\text{SD}}$ was between .200 and .400. The true standard deviation of $\hat{\theta}$ is .286 in this case, so the jackknife estimate is nearly unbiased, though quite variable, having coefficient of variation $.084/.280=.30$.

As a point of comparison, Table 3.1 also gives summary statistics for the bootstrap estimate of standard deviation introduced in Chapter 5. The bootstrap estimate is also nearly unbiased but with noticeably smaller variability from trial to trial. The figures in brackets show the minimum possible coefficient of variation in each case for any estimate of standard deviation which is scale invariant. In the normal case; for example, .19 is the coefficient of variation of $[\sum (x_i - \bar{x})^2/14]^{1/2}$.

3.4. The sample median. The trimmed mean with $\alpha \to .5$ is the sample median: $\hat{\theta} = x_{(m)}$ if $n = 2m-1$, $\hat{\theta} = (x_{(m)} + x_{(m+1)})/2$ if $n = 2m$. Formula (3.2) fails in this case. In the even case $n = 2m$, (3.2) gives

$$\widehat{\text{VAR}} = \frac{n-1}{4}[x_{(m+1)} - x_{(m)}]^2.$$

Standard asymptotic theory (Pyke (1965)) shows that if F has a density function f then

$$n\,\widehat{\text{VAR}} \xrightarrow{\mathcal{L}} \frac{1}{4f^2(\theta)}\left[\frac{\chi_2^2}{2}\right]^2$$

as $n \to \infty$, where $f(\theta)$ is the density at the true median θ, $f(\theta)$ assumed >0, and $[\chi_2^2/2]^2$ is a random variable with expectation 2, variance 20. The true variance of $\hat{\theta}$ goes to the limit

$$n \text{ Var} \to \frac{1}{4f^2(\theta)};$$

see Kendall and Stuart (1958). In this case $\widehat{\text{VAR}}$ is not even a consistent estimator of Var. An explanation of what goes wrong is given in Chapter 6. The bootstrap estimate of variance performs reasonably well for the sample median; see Chapter 10.

3.5. Ratio estimation. As in Example 2.3, $\mathscr{X} = \mathscr{R}^{2+}$, $X = (Y, Z)$, but here we consider the statistic $\hat{\theta} = \log \bar{z}/\bar{y}$. Table 3.2 reports the results of a Monte Carlo experiment: 100 trials of $(Y_1, Z_1), (Y_2, Z_2), \cdots, (Y_{10}, Z_{10}) \overset{\text{iid}}{\sim} F$. The two cases considered for F both had $Y \sim U(0, 1)$, the uniform distribution on [0, 1]. Z was taken independent of Y; in one case $Z \sim G_1$, in the other $Z \sim G_1^2/2$. The summary statistics for the 100 trials show that $\widehat{\text{SD}}$, the jackknife estimate of standard deviation, is nearly unbiased for the true Sd. Once again the estimates $\widehat{\text{SD}}$ are quite variable from trial to trial, which is perhaps not surprising given a sample size of only $n = 10$.

Table 3.2 also presents results for the bootstrap (Chapter 5) and the delta method (Chapter 6). The delta method is the best known of all the techniques we will discuss. In the present case it consists of approximating the Sd of $\hat{\theta} = \log \bar{z}/\bar{y}$ by the Sd of its first order Taylor series expansion, $\hat{\theta} \doteq \log(\mu_z/\mu_y) + (\bar{z} - \mu_z)/\mu_z - (\bar{y} - \mu_y)/\mu_y$, where $\mu_z = E_F Z$, $\mu_y = E_F Y$. The resulting approximation, $\text{Sd}(\hat{\theta}) \doteq [\mu_{yy}/\mu_y^2 + \mu_{zz}/\mu_z^2 - 2\mu_{yz}/\mu_y\mu_z]/n$, where $\mu_{yy} = E_F[Y - EF_Y]^2$, etc., is then estimated by substituting sample moments $\hat{\mu}_y$, $\hat{\mu}_z$,

TABLE 3.2

Estimates of standard deviation for $\hat{\theta} = \log \bar{z}/\bar{y}$; 100 trials, sample size $n = 10$ for each trial. Summary statistics of $\widehat{\text{SD}}$ for the 100 trials show that the jackknife is more variable than the bootstrap or the delta method. However the delta method is badly biased in the second case.

	$Y \sim U(0,1), Z \sim G_1$			$Y \sim U(0,1), Z \sim G_1^2/2$		
	Ave	Std dev	Coeff var	Ave	Std dev	Coeff var
Jackknife	.37	.11	.30	.70	.33	.47
Bootstrap, 1000 reps per trial	.37	.10	.27	.64	.23	.36
Delta method	.35	.09	.26	.53	.14	.26
True Sd	.37			.67		

$\hat{\mu}_{yy}, \hat{\mu}_{zz}, \hat{\mu}_{yz}$ for the unknown true population moments. In the present situation the delta method has the lowest coefficient of variation, but is badly biased downwards in the second case. The delta method is discussed further in Chapter 6.

3.6. Functions of the expectation. Suppose $\mathscr{X} = \mathscr{R}^1$, $\hat{\theta} = g(\bar{x})$, where g is some nicely behaved function such as $\sin(\bar{x})$ or $1/(1+\bar{x})$. (For what follows we need that the derivative g' exists continuously.) Then a first order Taylor series expansion gives

$$\hat{\theta}_{(i)} = g\left(\frac{n\bar{x} - x_i}{n-1}\right) \doteq g(\bar{x}) + g'(\bar{x})\frac{\bar{x} - x_i}{n-1},$$

so, substituting the expansion into (3.2), we have

$$\text{(3.9)} \qquad \widehat{\text{VAR}} \doteq \frac{n-1}{n}[g'(\bar{x})]^2 \frac{1}{(n-1)^2} \sum (x_i - \bar{x})^2 = [g'(\bar{x})]^2 \frac{\hat{\sigma}^2}{n},$$

where $\hat{\sigma}^2 = \sum (x_i - \bar{x})^2/(n-1)$ is the usual unbiased estimate of variance.

The variance of $\hat{\theta} = g(\bar{x})$ is usually obtained by the delta method: $g(\bar{x}) \doteq g(\mu) + g'(\mu)(\bar{x} - \mu)$, where $\mu = E_F X$, so $\text{Var}\{\hat{\theta}\} \doteq [g'(\mu)]^2 \text{Var}\{\bar{X}\}$. Estimating μ by $\bar{x}$ and $\text{Var}\{\bar{X}\}$ by $\hat{\sigma}^2/n$ gives (3.9). We have shown that the delta method gives the same estimate of variance as the jackknife, if a linear approximation is used to simplify the latter. Chapter 6 discusses the "infinitesimal jackknife", a variant of the jackknife which gives *exactly* the same estimates as the delta method, for both bias and variance, whenever the delta method applies.

3.7. The law school data. For the correlation coefficient $\hat{\rho}$ based on the law school data (Table 2.2 and Fig. 2.2) we calculate $\widehat{\text{SD}} = .142$. This might be compared with the normal theory estimate $(1-\hat{\rho}^2)/\sqrt{n-3} = (1-.776^2)/\sqrt{12} = .115$; see Johnson and Kotz (1970, p. 229). A disturbing feature of $\widehat{\text{VAR}}$ can be seen in Table 2.2: 55% of $\sum [\hat{\rho}_{(i)} - \hat{\rho}_{(\cdot)}]^2$ comes from data point 1. $\widehat{\text{VAR}}$ is not robust in the case of $\hat{\rho}$, a point discussed in Hinkley (1978).

TABLE 3.3

Estimates of standard deviation for $\hat{\rho}$, the correlation coefficient, and $\tanh^{-1}\hat{\rho}$: 200 trials, $X_1, X_2, \cdots, X_{14} \overset{\text{iid}}{\sim} F$, F bivariate normal with true $\rho = .5$. The jackknife is more variable than the bootstrap or the delta method, but the latter is badly biased downward.

	$\hat{\rho}$			$\tanh^{-1}\hat{\rho}$		
	Ave	Std dev	Coeff var	Ave	Std dev	Coeff var
Jackknife	.223	.085	.38	.314	.090	.29
Bootstrap, 512 reps per trial	.206	.063	.31	.301	.062	.21
Delta method	.175	.058	.33	.244	.052	.21
True Sd	.218			.299		

Table 3.3 reports the results of a Monte Carlo experiment: 200 trials of $X_1, X_2, \cdots, X_{14} \sim F$ bivariate normal, true $\rho = .50$. Two statistics were considered, $\hat{\rho}$ and Fisher's transformation $\tanh^{-1}\hat{\rho} = \frac{1}{2}\log(1+\hat{\rho})/(1-\hat{\rho})$. In this case the jackknife $\widehat{\text{SD}}$ is considerably more variable than the bootstrap $\widehat{\text{SD}}$.

3.8. Linear regression. Consider the linear regression model

$$y_i = c_i\beta + \varepsilon_i, \qquad i = 1, 2, \cdots, n,$$

where $\varepsilon_i \overset{\text{iid}}{\sim} F$, F an unknown distribution on $\mathcal{R}^1$ having $E_F\varepsilon = 0$. Here c_i is a known $1 \times p$ vector of covariates while β is a $p \times 1$ vector of unknown parameters. The statistic of interest is the least squares estimate of β,

$$\hat{\beta} = \mathbf{G}^{-1}\mathbf{C}'\mathbf{y}, \tag{3.10}$$

$\mathbf{y} = (y_1, y_2, \cdots, y_n)'$, $\mathbf{C}' = (c_1', c_2', \cdots, c_n')$, $\mathbf{G} = \mathbf{C}'\mathbf{C}$. We assume that the $p \times n$ matrix $\mathbf{C}$ is nonsingular, so that the $p \times p$ matrix $\mathbf{G}$ has an inverse.

The usual estimate of Cov $(\hat{\beta})$, the covariance matrix of $\hat{\beta}$, is

$$\hat{\sigma}^2\mathbf{G}^{-1}, \qquad \hat{\sigma}^2 = \sum_{i=1}^{n} \frac{\hat{\varepsilon}_i^2}{n-p}, \tag{3.11}$$

$\hat{\varepsilon}_i$ being the estimated residual $y_i - c_i\hat{\beta}$. We don't need the jackknife in this situation, but it is interesting to compare $\widehat{\text{VAR}}$ with (3.11).

The statistic $\hat{\beta}$ is not a symmetric function of $y_1, y_2, \cdots, y_n$, but it is symmetrically defined in terms of the vectors $(c_1, y_1), (c_2, y_2), \cdots, (c_n, y_n)$. What we have called x_i before is now the $p+1$ vector (c_i, y_i). Let $\hat{\beta}_{(i)}$ be the statistic (3.10) computed with (c_i, y_i) removed, which turns out to be

$$\hat{\beta}_{(i)} = \hat{\beta} - \frac{\mathbf{G}^{-1}c_i\hat{\varepsilon}_i}{1 - c_i\mathbf{G}^{-1}c_i'}; \tag{3.12}$$

see Miller (1974b), Hinkley (1977). The multivariate version of Tukey's formula (3.2) is

$$(3.13)\qquad \widehat{\mathrm{COV}}=\frac{n-1}{n}\sum_{i=1}^{n}[\hat{\beta}_{(i)}-\hat{\beta}_{(\cdot)}][\hat{\beta}_{(i)}-\hat{\beta}_{(\cdot)}]' \doteq \frac{n-1}{n}\mathbf{G}^{-1}\left[\sum c_i'c_i\hat{\varepsilon}_i^2\right]\mathbf{G}^{-1}.$$

This last formula ignores the factor $1-c_i\mathbf{G}^{-1}c_i'=1-O(1/n)$ in the denominator of (3.12), which turns out to be equivalent to using the infinitesimal jackknife–delta method.

If all the $\hat{\varepsilon}_i^2$ are identical in value then (3.13) is about the same as the standard answer (3.11), but otherwise the two formulas are quite different. We will see why when we apply the bootstrap to regression problems in Chapter 5.

CHAPTER 4

Bias of the Jackknife Variance Estimate

This chapter shows that the jackknife variance estimate tends to be conservative in the sense that its expectation is greater than the true variance. The actual statement of the main theorem given below is necessarily somewhat different, but all of our Monte Carlo results, for example Tables 3.1–3.3, confirm that $\widehat{\mathrm{VAR}}$ is, if anything, biased moderately upward.[1] This contrasts with the delta method, which we have seen to be capable of severe downward biases.

The material of this chapter is somewhat technical, and can be skipped by readers anxious to get on with the main story. On the other hand, it is nice to have a precise result in the midst of so much approximation and heuristic reasoning. A fuller account of these results is given in Efron and Stein (1981). Section 4.3, concerning influence functions, is referred to in Chapter 6.

Once again, let $\hat{\theta}(X_1, X_2, \cdots, X_n)$ be a statistic symmetrically defined in its n arguments, these being an i.i.d. sample from an unknown distribution F on an arbitrary space $\mathscr{X}$, $X_1, X_2, \cdots, X_n \overset{\text{iid}}{\sim} F$. In order to use the jackknife formula (3.2), it is necessary that $\hat{\theta}$ also be defined for sample size $n-1$. Let Var_n be the variance of $\hat{\theta}(X_1, X_2, \cdots, X_n)$, Var_{n-1} the variance of $\hat{\theta}(X_1, X_2, \cdots, X_{n-1})$, and define

$$(4.1)\qquad \widetilde{\mathrm{VAR}} = \sum_{i=1}^{n} [\hat{\theta}_{(i)} - \hat{\theta}_{(\cdot)}]^2.$$

It is useful to think of $\widehat{\mathrm{VAR}}$, formula (3.2), as estimating the true variance Var_n in two distinct steps:

(i) a direct estimate of Var_{n-1}, namely $\widetilde{\mathrm{VAR}}$, and

(ii) a sample size modification to go from $n-1$ to n,

$$(4.2)\qquad \widehat{\mathrm{VAR}} = \frac{n-1}{n} \widetilde{\mathrm{VAR}}.$$

The main result of this chapter concerns (i). We show that

$$(4.3)\qquad E_F\, \widetilde{\mathrm{VAR}} \geqq \mathrm{Var}_{n-1}.$$

$\widetilde{\mathrm{VAR}}$ always overestimates Var_{n-1} in expectation. We also discuss, briefly, the sample size modification (ii), which is based on the fact that for many familiar

[1] Notice that in Table 3.1, F normal, the 200 jackknife estimates $\widehat{\mathrm{VAR}}$ averaged .0854 $(=.280^2 + (.199/200).084^2)$ compared to the true variance .0816 $(=.286^2)$.

statistics

$$\text{Var}_n = \frac{n-1}{n}\text{Var}_{n-1} + O\left(\frac{1}{n^3}\right), \tag{4.4}$$

the $O(1/n^3)$ term being negligible compared to the modification $(1/n)\,\text{Var}_{n-1} = O(1/n^2)$. For $\hat\theta = \bar x$, $\text{Var}_n = ((n-1)/n)\,\text{Var}_{n-1}$ exactly.

4.1. ANOVA decomposition of $\hat\theta$. This will be the main tool in proving (4.3). It is a decomposition of $\hat\theta(X_1, X_2, \cdots, X_n)$ based on the ANOVA decomposition of a complete n-way table. Assume that $E_F\hat\theta^2 < \infty$, and define

$$\begin{aligned} &\mu = E_F\hat\theta,\\ &\alpha_i = \alpha(X_i) = n[E_F\{\hat\theta|X_i\} - \mu],\\ &\beta_{ii'} = \beta(X_i, X_{i'}) = n^2[E_F\{\hat\theta|X_i, X_{i'}\} - E_F\{\hat\theta|X_i\} - E_F\{\hat\theta|X_{i'}\} + \mu] \quad \text{for } i \neq i', \end{aligned} \tag{4.5}$$

etc., the last definition being

$$\begin{aligned} \eta_{123\cdots n} &= \eta(X_1, X_2, X_3, \cdots, X_n)\\ &= n^n[\hat\theta - E_F\{\hat\theta|X_1, X_2, \cdots, X_{n-1}\}\\ &\qquad - E_F\{\hat\theta|X_1, X_2, \cdots, X_{n-2}, X_n\} - \cdots + (-1)^n\mu]. \end{aligned}$$

In the usual ANOVA terminology μ is the grand mean, $n\alpha_i$ is the ith main effect, $n^2\beta_{ii'}$ is the ii'th interaction, etc. The reason for multiplying by powers of n is discussed below.

There are 2^n random variables $\mu, \alpha_i, \beta_{ii'}, \cdots, \eta_{12\cdots n}$ defined above, corresponding to the 2^n possible subsets of $\{1, 2, \cdots, n\}$. They have three salient properties:

Property 1. Each random variable is a function only of the X_i indicated by its subscripts (e.g., β_{37} is a function of X_3 and X_7).

Property 2. Each random variable has conditional expectation 0, when conditioned upon all but one of its defining X_i (e.g., $E_F\alpha_1 = 0$, $E_F\{\beta_{12}|X_2\} = 0$).

Property 3. $\hat\theta$ decomposes into a sum of $\mu, \alpha_i, \beta_{ii'}, \cdots, \eta_{123\cdots n}$ as follows:

$$\hat\theta(X_1, X_2, \cdots, X_n) = \mu + \frac{1}{n}\sum_i \alpha_i + \frac{1}{n^2}\sum_{i<i'} \beta_{ii'} + \cdots + \frac{1}{n^n}\eta_{123\cdots n}. \tag{4.6}$$

($i < i'$ is short for $1 \leqq i < i' \leqq n$, etc.)

The proofs of Properties 2 and 3 are essentially the same as those for the standard ANOVA decomposition of an n-way table (see Scheffe (1959, § 4.5)) and are given in Efron and Stein (1981). Notice that Property 2 implies

Property 2′. The 2^n random variables $\mu, \alpha_i, \beta_{ii'}, \cdots, \eta_{123\cdots n}$ are mutually uncorrelated (e.g., $E_F\alpha_1\alpha_2 = 0$, $E_F\alpha_1\beta_{12} = 0$).

4.2. Proof of the main result. First of all, notice that the main result (4.3) concerns only samples of size $n-1$. It is really a statement about $\hat\theta(X_1, X_2, \cdots, X_{n-1})$, and there is no need to define the original statistic

$\hat{\theta}(X_1, X_2, \cdots, X_n)$. We will need both $\hat{\theta}(X_1, X_2, \cdots, X_{n-1})$ and $\hat{\theta}(X_1, X_2, \cdots, X_n)$ when we discuss (4.4).

Consider decomposition (4.6) for $\hat{\theta}(X_1, X_2, \cdots, X_{n-1})$, and define

$$\sigma_\alpha^2 = \text{Var}_F\, \alpha_i, \qquad \sigma_\beta^2 = \text{Var}_F\, \beta_{ii'}, \quad \text{etc.} \tag{4.7}$$

Using Property 2′, we can immediately calculate $\text{Var}\{\hat{\theta}(X_1, X_2, \cdots, X_n)\} = \text{Var}_{n-1}$ simply by counting the terms in (4.6),

$$\text{Var}_{n-1} = \frac{\sigma_\alpha^2}{n-1} + \binom{n-2}{1} \frac{\sigma_\beta^2}{2(n-1)^3} + \binom{n-2}{2} \frac{\sigma_\gamma^2}{3(n-1)^5} + \cdots. \tag{4.8}$$

(Remember we are applying (4.6) to $\hat{\theta}(X_1, \cdots, X_{n-1})$, so, for example, the α term is $\sum_{i=1}^{n-1} \alpha_i/(n-1)$.) There are $n-1$ terms on the right side of (4.8).

The statistic $\hat{\theta}(X_1, X_2, \cdots, X_{n-1})$ is what we have previously called $\hat{\theta}_{(n)}$. We can also apply (4.6) to $\hat{\theta}_{(n-1)} = \hat{\theta}(X_1, X_2, \cdots, X_{n-2}, X_n)$, take the difference and calculate

$$E_F[\hat{\theta}_{(n)} - \hat{\theta}_{(n-1)}]^2 = 2\left(\frac{\sigma_\alpha^2}{(n-1)^2} + \binom{n-2}{1} \frac{\sigma_\beta^2}{(n-1)^4} + \cdots\right). \tag{4.9}$$

Since

$$\widetilde{\text{VAR}} = \sum_{i=1}^{n} [\hat{\theta}_{(i)} - \hat{\theta}_{(\cdot)}]^2 = \frac{1}{n} \sum_{1 \le i < i' \le n} [\hat{\theta}_{(i)} - \hat{\theta}_{(i')}]^2,$$

and all the terms in this last expression have expected value (4.9), we get

$$E_F\, \widetilde{\text{VAR}} = \frac{\sigma_\alpha^2}{n-1} + \binom{n-2}{1} \frac{\sigma_\beta^2}{(n-1)^3} + \binom{n-2}{2} \frac{\sigma_\gamma^2}{(n-1)^5} + \cdots. \tag{4.10}$$

Comparing (4.8) with (4.10) gives

THEOREM 4.1. *$E_F\, \widetilde{\text{VAR}}$ exceeds Var_{n-1} by an amount*

$$E_F\, \widetilde{\text{VAR}} - \text{Var}_{n-1} = \binom{n-2}{1} \frac{\sigma_\beta^2}{2(n-1)^3} + 2\binom{n-2}{2} \frac{\sigma_\gamma^2}{3(n-1)^5} + \cdots, \tag{4.11}$$

there being $n-2$ terms on the right side of (4.11).

Several comments are pertinent. (1) All the terms on the right side of (4.11) are positive, so this proves the main result (4.3). (2) A *linear functional* is by definition a statistic of the form

$$\hat{\theta} = \mu + \frac{1}{n} \sum_i \alpha(X_i). \tag{4.12}$$

If $\hat{\theta}(X_1, X_2, \cdots, X_{n-1})$ is linear, the right side of (4.11) is zero and $E_F\, \widetilde{\text{VAR}} = \text{Var}_{n-1}$; otherwise $E_F\, \widetilde{\text{VAR}} > \text{Var}_{n-1}$. (3) Comparing (4.10) with (4.8) shows that $E_F\, \widetilde{\text{VAR}}$ doubles the quadratic (σ_β^2) term in Var_{n-1}, triples the cubic term, etc. If $\widetilde{\text{VAR}}$ is not to be badly biased upwards, then most of the variance of $\hat{\theta}(X_1, \cdots, X_{n-1})$ must be contained in the linear term $\sum_i \alpha_i/(n-1)$. This seems usually to be the case. The theory of influence functions discussed below shows that it is asymptotically true under suitable regularity conditions on $\hat{\theta}$. (4) Tables 3.1–3.3 allow us to estimate the proportion of the variance of $\hat{\theta}(X_1, X_2, \cdots, X_{n-1})$ contained in the linear term. The estimated proportions are

Table 3.1: 96%, 93%,

Table 3.2: 91%, 67%,

Table 3.3: 83%, 81%.

(For example, $.96 = 1-(.0854-.0816)/.0816$; see the footnote at the beginning of this chapter.) (5)A version of (4.3) applicable to any function $S(X_1, X_2, \cdots, X_{n-1})$, symmetric or not, of any $n-1$ independent arguments, identically distributed or not, is given in Efron and Stein (1981). (6) The terms σ_α^2, σ_β^2, etc. depend on the sample size; see the discussion of quadratic functionals below. (7) Steele (1980) has used (4.3) in the proof of certain conjectures concerning subadditive functionals in the plane.

4.3. Influence functions. The influence function IF (x) of a functional statistic $\hat{\theta} = \theta(\hat{F})$ evaluated at the true probability distribution F, is defined as

$$\text{IF}(x) = \lim_{\varepsilon \to 0} \frac{\theta((1-\varepsilon)F + \varepsilon\delta_x) - \theta(F)}{\varepsilon},$$

where δ_x is a unit probability mass on the point x. Under reasonable conditions (see Huber (1974) or Hampel (1974))

$$\theta(\hat{F}) = \theta(F) + \frac{1}{n}\sum_{i=1}^{n} \text{IF}(X_i) + O_p\left(\frac{1}{n}\right). \tag{4.13}$$

This formula looks like the beginning of (4.6).

As a matter of fact, the function $\alpha(x)$ converges to IF (x) as $n \to \infty$, again under suitable regularity conditions. That is the reason for multiplying by n in (4.5). Likewise $\beta(x, x')$ converges to the second order influence function, etc. In a sense $\alpha(x)$ deserves to be called the finite sample influence function, a point discussed by Mallows (1974).

4.4. Quadratic functionals. Consider the statistic $\hat{\theta}(X_1, X_2, \cdots, X_n) = \sum_{i=1}^{n} [X_i - \bar{X}]^2/n$, where the X_i are i.i.d. with expectation ξ and variance σ^2, $X_i \overset{\text{iid}}{\sim} (\xi, \sigma^2)$. Expansion (4.6) can be evaluated explicitly in this case:

$$\hat{\theta}(X_1, X_2, \cdots, X_n) = \mu^{(n)} + \frac{1}{n}\sum_i \alpha^{(n)}(X_i) + \frac{1}{n^2}\sum_{i<i'} \beta(X_i, X_{i'}), \tag{4.14}$$

where

$$\mu^{(n)} = \frac{n-1}{n}\sigma^2, \qquad \alpha^{(n)}(x) = \frac{n-1}{n}[(x-\xi)^2 - \sigma^2],$$
$$\beta(x, x') = -2(x-\xi)(x'-\xi). \tag{4.15}$$

This is an example of a *quadratic functional*: the ANOVA decomposition (4.14) terminates at the quadratic term, and $\hat{\theta}$ is a functional statistic, $\hat{\theta} = \theta(\hat{F})$.

Notice in (4.15) that $\beta(\cdot, \cdot)$ does not depend upon n, while $\mu^{(n)}$ and $\alpha^{(n)}(\cdot)$ both involve $1/n$ terms. ($\lim_{n\to\infty} \alpha^{(n)}(x) = \alpha^{(\infty)}(x) = (x-\xi)^2 + \sigma^2$, the influence function of $\hat{\theta}$, as mentioned above; $\lim_{n\to\infty} \mu^{(n)} = \mu^{(\infty)} = \theta(F)$.) This turns out to be true in general: $\hat{\theta}$ is a quadratic functional if and only if it can be written in form (4.14) with

$$(4.16)\quad \mu^{(n)} = \mu^{(\infty)} + \frac{E_F\beta(X, X)}{2n}, \qquad \alpha^{(n)}(x) = \alpha^{(\infty)}(x) + \frac{\beta(x, x) - E_F\beta(X, X)}{2n};$$

see Efron and Stein (1981).

Quadratic functionals are the simplest nonlinear functional statistics and as such they can be used to understand the problems caused by nonlinearity, as demonstrated at the end of this chapter. Theorem 2.1 shows that they justify the jackknife estimate of bias. We can now prove Theorem 2.1. From (4.14) we get

$$(4.17)\qquad \hat{\theta}_{(\cdot)} = \mu^{(n-1)} + \frac{1}{n}\sum_{1\le i\le n} \alpha_i^{(n-1)} + \frac{1}{n^2}\frac{n(n-2)}{(n-1)^2}\sum_{1\le i<i'\le n} \beta_{ii'}.$$

Define

$$(4.18)\qquad \Delta_i = \Delta(X_i) = \frac{\beta(X_i, X_i)}{2}, \qquad E_F\Delta = E_F\Delta(X).$$

Then (4.16) can be rewritten as

$$(4.19)\qquad \mu^{(n-1)} - \mu^{(n)} = \frac{E_F\Delta}{n(n-1)}, \qquad \alpha^{(n-1)}(x) - \alpha^{(n)}(x) = \frac{\Delta(x) - E_F\Delta}{n(n-1)}.$$

Subtracting (4.14) from (4.17) gives

$$\widehat{\text{BIAS}} = (n-1)(\hat{\theta}_{(\cdot)} - \hat{\theta}) = (n-1)[\mu^{(n-1)} - \mu^{(n)}]$$

$$+\frac{n-1}{n}\sum_i [\alpha_i^{(n-1)} - \alpha_i^{(n)}] + \frac{1}{n^2}\left[\frac{n(n-2)}{n-1} - (n-1)\right]\sum_{i<i'} \beta_{ii'}$$

$$= \frac{E\Delta}{n} + \frac{1}{n}\sum_i \frac{\Delta_i - E_F\Delta}{n} - \frac{1}{n^2(n-1)^2}\sum_{i<i'} \beta_{ii'}.$$

Taking the expectation of this last expression gives

$$E_F\,\widehat{\text{BIAS}} = \frac{E_F\Delta}{n}$$

since $\Delta_i - E_F\Delta$ and $\beta_{ii'}$ all have expectation zero. However, the first equation in

(4.16) shows that

$$\frac{E_F\Delta}{n} = \mu^{(n)} - \mu^{(\infty)} = E_F\theta(\hat{F}) - \theta(F),$$

which is the statement of Theorem 2.1.

4.5. Sample size modification. It is *not* always true that $E_F\ \widehat{\text{VAR}} \geqq \text{Var}_n$. Arbitrarily bad counterexamples can be constructed, which is not surprising since $\widehat{\text{VAR}}$ is defined entirely in terms of samples of size $n-1$ while Var_n is the variance for sample size n. However, for many reasonable classes of statistics we do have $E_F\ \widehat{\text{VAR}} \geqq \text{Var}_n$, either asymptotically or for all n. Three examples are discussed in Efron and Stein (1981).

(1) *U-statistics.* A U-statistic is a statistic of the form

$$\hat{\theta}(X_1, X_2, \cdots, X_n) = \sum_{i_1 < i_2 < \cdots < i_k} \frac{g(X_{i_1}, X_{i_2}, \cdots, X_{i_k})}{\binom{n}{k}},$$

with k some fixed integer and g some fixed symmetric function of k arguments. Hoeffding (1948) showed that, for $n-1 \geqq k$, the smallest possible sample size, U-statistics satisfy $((n-1)/n)\,\text{Var}_{n-1} \geqq \text{Var}_n$. Combining this with Theorem 4.1 and definition (4.2) gives

$$E_F\ \widehat{\text{VAR}} = \frac{n-1}{n} E_F\ \widetilde{\text{VAR}} \geqq \frac{n-1}{n} \text{Var}_{n-1} \geqq \text{Var}_n,$$

the desired result. *Note.* Hoeffding's important paper uses what we have called the ANOVA decomposition, although in rearranged form.

(2) *Von Mises series.* This is a term coined by C. Mallows for statistics of the form

$$\hat{\theta}(X_1, X_2, \cdots, X_n) = \mu + \frac{1}{n}\sum_i \alpha(X_i) + \frac{1}{n^2}\sum_{i<i'} \beta(X_i, X_{i'}) + \cdots$$

$$+ \frac{1}{n^k} \sum_{i_1 < i_2 < \cdots < i_k} \phi(X_{i_1}, X_{i_2}, \cdots, X_{i_k}).$$

Here k is a fixed integer, $\mu, \alpha, \beta, \cdots, \phi$ are fixed functions not depending on n and $n \geqq k$. A von Mises series is not quite a U-statistic. (It becomes one if we divide by $n, n(n-1), \cdots$ instead of $n, n^2, \cdots$ in the definition.) Efron and Stein (1981) show that in this case it is *not* always true that $(n-1)/n\ \text{Var}_{n-1} \geqq \text{Var}_n$, but it is still true that $E_F\ \text{VAR} \geqq \text{Var}_n$, the desired result.

(3) *Quadratic functionals.* For a quadratic functional,

$$\text{(4.20)} \quad E_F\ \widehat{\text{VAR}} = \text{Var}_n + \frac{1}{n(n-1)}\left\{\frac{n^3 - n^2 - 3n + 1}{n^3 - n^2}\frac{\sigma_\beta^2}{2} + \frac{2\sigma_{\alpha\Delta}}{n} + \frac{\sigma_\Delta^2}{n^2(n-1)}\right\},$$

where $\sigma_{\alpha\Delta} = E_F\alpha(X)\Delta(X)$, $\sigma_\Delta^2 = E_F[\Delta(X) - E_F\Delta]^2$. By constructing examples with $\sigma_{\alpha\Delta}$ sufficiently negative, we can force $E_F\ \widehat{\text{VAR}} < \text{Var}_n$ for small n. For n large, a sufficient condition being $n > -4\sigma_{\alpha\Delta}/\sigma_\beta^2$, we again have $E_F\ \widehat{\text{VAR}} \geqq \text{Var}_n$.

CHAPTER 5

The Bootstrap

The bootstrap (Efron (1979a)) is conceptually the simplest of all the techniques considered here. We begin our discussion with the bootstrap estimate of standard deviation, which performed well in Tables 3.1–3.3, and then go on to more general problems. The connection with the jackknife is made in Chapter 6.

Given a statistic $\hat{\theta}(X_1, X_2, \cdots, X_n)$ defined symmetrically in $X_1, X_2, \cdots, X_n \overset{\text{iid}}{\sim} F$, write the standard deviation of $\hat{\theta}$ as

$$\text{Sd} = \sigma(F, n, \hat{\theta}) = \sigma(F). \tag{5.1}$$

This last notation emphasizes that, given the sample size n and the form of the statistic $\hat{\theta}(\cdot, \cdot, \cdot, \cdot)$, the standard deviation is a function of the unknown probability distribution F. The bootstrap estimate of standard deviation is simply $\sigma(\cdot)$ evaluated at $F = \hat{F}$,

$$\widehat{\text{SD}} = \sigma(\hat{F}). \tag{5.2}$$

Since $\hat{F}$ is the nonparametric maximum likelihood estimate of F, another way to express (5.2) is that $\widehat{\text{SD}}$ is the nonparametric MLE of Sd.

Example 5.1. *The average.* $\mathscr{X} = \mathscr{R}^1$ and $\hat{\theta}(X_1, X_2, \cdots, X_n) = \bar{X}$. In this case we know that the standard deviation of $\bar{X}$ is

$$\sigma(F) = \left[\frac{\mu_2}{n}\right]^{1/2},$$

where $\mu_2 = E_F[X - E_F X]^2$, the second central moment of F, as in (3.5). Therefore

$$\widehat{\text{SD}} = \sigma(\hat{F}) = \left[\frac{\hat{\mu}_2}{n}\right]^{1/2}, \tag{5.3}$$

$\hat{\mu}_2 = \sum_{i=1}^{n} [x_i - \bar{x}]^2/n$ being the second central moment of $\hat{F}$, i.e., the sample value of μ_2. This is not quite the usual estimate of Sd since it uses the maximum likelihood rather than the unbiased estimate of μ_2. The bootstrap variance estimate $\widehat{\text{VAR}} = \hat{\mu}_2/n$ is biased downward,

$$E_F\,\widehat{\text{VAR}} = E_F\left[\frac{\hat{\mu}_2}{n}\right] = \frac{n-1}{n}\frac{\mu_2}{n} = \frac{n-1}{n}\,\text{Var}\,\{\bar{X}\}.$$

We could rescale (5.2) to make $\widehat{\text{VAR}}$ unbiased in the case $\hat{\theta} = \bar{X}$, i.e. define $\widehat{\text{SD}} = [n/(n-1)]^{1/2}\sigma(\hat{F})$, but this doesn't seem to give better Sd estimates. The jackknife estimate of standard deviation *is* rescaled in this way, as will be made clear in § 5.6.

5.1. Monte Carlo evaluation of $\widehat{\mathrm{SD}}$. Usually the function $\sigma(F)$ cannot be written down explicitly. In order to carry out the calculation of $\widehat{\mathrm{SD}}$, (5.2), it is then necessary to use a Monte Carlo algorithm.

1. Fit the nonparametric MLE of F,

$$\hat{F}: \text{mass } \frac{1}{n} \text{ at } x_i, \qquad i=1,2,\cdots,n. \tag{5.4}$$

2. Draw a "bootstrap sample" from $\hat{F}$,

$$X_1^*, X_2^*, \cdots, X_n^* \overset{\text{iid}}{\sim} \hat{F}, \tag{5.5}$$

and calculate $\hat{\theta}^* = \hat{\theta}(X_1^*, X_2^*, \cdots, X_n^*)$.

3. Independently repeat step 2 a large number B of times, obtaining "bootstrap replications" $\hat{\theta}^{*1}, \hat{\theta}^{*2}, \cdots, \hat{\theta}^{*B}$, and calculate

$$\widehat{\mathrm{SD}} = \left\{ \frac{1}{B-1} \sum_{b=1}^{B} [\hat{\theta}^{*b} - \hat{\theta}^{*\cdot}]^2 \right\}^{1/2}. \tag{5.6}$$

The dot notation indicates averaging: $\hat{\theta}^{*\cdot} = \sum_{b=1}^{B} \hat{\theta}^{*b}/B$.

If we could let $B\to\infty$ then (5.6) would exactly equal (5.2). In practice we have to stop the bootstrap process sooner or later, sooner being preferable in terms of computational cost. Tables 3.1, 3.2, 3.3 used $B=200$, 1000, 512 respectively. These values were deliberately taken large for investigating quantities other than $\widehat{\mathrm{SD}}$ and $B=100$ performed almost as well in all three situations. In most cases there is no point in taking B so large that (5.6) agrees very closely with (5.2), since (5.2) itself will be highly variable for estimating the true Sd. This point will be discussed further as we go through the examples.

Example 5.2. *Switzer's adaptive trimmed mean.* The charm of the jackknife and the bootstrap is that they can be applied to complicated situations where parametric modeling and/or theoretical analysis is hopeless. As a relatively simple "complicated situation", consider Switzer's adaptive trimmed mean $\hat{\theta}(x_1, x_2, \cdots, x_n)$, defined in Carroll (1979):

(i) given the data $x_1, x_2, \cdots, x_n$, compute the jackknife estimate of variance for the 5%, 10% and 25% trimmed means, and

(ii) let $\hat{\theta}$ be the value of the trimmed mean corresponding to the minimum of the three variance estimates.

The results of a large Monte Carlo study are shown in Table 5.1. Two sample sizes, $n=10$, 20, and three distributions, $F\sim\mathcal{N}(0,1)$, G_1, $e^{\mathcal{N}(0,1)}$, were investigated. $B=200$ bootstrap replications were taken for each trial.[1] The bootstrap clearly outperforms the jackknife, except for the case $n=10$, $F\sim e^{\mathcal{N}(0,1)}$, for which both are ineffective. The bootstrap results are surprisingly close to the theoretical optimum for a scale invariant Sd estimator, assuming full knowledge of the parametric family, when $F\sim\mathcal{N}(0,1)$ and $F\sim G_1$.

[1] "Trial" always refers to a new selection of the original data $X_1, X_2, \cdots, X_n \sim F$, while "replication" refers to a selection of the bootstrap data $X_1^*, X_2^*, \cdots, X_n^* \sim \hat{F}$.

TABLE 5.1

Estimates of standard deviation for Switzer's adaptive trimmed mean using the jackknife and the bootstrap. The minimum possible Coefficient of Variation for a scale invariant estimate of standard deviation, assuming full knowledge of the parametric family, is shown for $F \sim \mathcal{N}(0,1)$ and G_1.

		Sample size $n = 10$			Sample size $n = 20$		
		$F \sim \mathcal{N}(0,1)$	G_1	$e^{\mathcal{N}(0,1)}$	$\mathcal{N}(0,1)$	G_1	$e^{\mathcal{N}(0,1)}$
Jackknife $\widehat{\text{SD}}$	Ave	.327	.296	.421	.236	.234	.324
	Std dev	.127	.173	.335	.070	.143	.228
	[Coeff var]	[.39]	[.58]	[.80]	[.30]	[.61]	[.70]
Bootstrap $\widehat{\text{SD}}$ $B = 200$	Ave	.328	.310	.541	.236	.222	.339
	Std dev	.081	.123	.310	.047	.072	.142
	[Coeff var]	[.25]	[.40]	[.57]	[.20]	[.32]	[.42]
True Sd		.336	.306	.483	.224	.222	.317
[Min possible Cv]		[.24]	[.33]		[.16]	[.23]	
No. of trials		1000	3000	1000	1000	3000	1000

Example 5.3. *The law school data.* Again referring to Table 2.1 and Fig. 2.2, $B = 1000$ bootstrap replications of the correlation coefficient were generated. Each of the 1000 bootstrap samples consisted of drawing 15 points *with* replacement from the 15 original data points shown in Fig. 2.2, the corresponding bootstrap replication being the correlation coefficient of the resampled points. A typical bootstrap sample might consist of law school 1 twice, law school 2 zero times, law school 3 once, etc. (Notice that the expected proportion of points in the original sample absent from the bootstrap sample is $(1-\frac{1}{15})^{15} = .36 \doteq e^{-1}$.)

The bootstrap estimate of $\widehat{\text{SD}}$, (5.6) was equal to .127, intermediate between the normal theory estimate .115 and the jackknife estimate .142. Figure 5.1 displays the histogram of the 1000 differences $\hat{\rho}^{*b} - \hat{\rho}$. Also shown is the normal theory density of $\hat{\rho}^*$, centered at $\hat{\rho}$, if the true correlation coefficient is $\rho = \hat{\rho} = .776$ (Johnson and Kotz (1970, p. 222)). The similarity between the histogram and the density curve suggests that the bootstrap replications may contain information beyond that used in calculating (5.6). We consider this point in Chapter 10, where we try to construct nonparametric confidence intervals.

5.2. Parametric bootstrap. Fisher's familiar theory for assigning a standard error to a maximum likelihood estimate is itself a "bootstrap theory", carried out in a parametric framework. Consider the law school example again, and suppose we are willing to accept a bivariate normal model for the data. The parametric maximum likelihood estimate of the unknown F is

(5.7) $$\hat{F}_{\text{NORM}} \approx \mathcal{N}_2\left(\begin{pmatrix}\hat{\mu}_y \\ \hat{\mu}_z\end{pmatrix}, \begin{pmatrix}\hat{\mu}_{yy} & \hat{\mu}_{yz} \\ \hat{\mu}_{yz} & \hat{\mu}_{zz}\end{pmatrix}\right),$$

where, if $X = (Y, Z)$ denotes a typical bivariate data point, $\hat{\mu}_y = \bar{y}$, $\hat{\mu}_{yz} = \sum (y_i - \bar{y})(z_i - \bar{z})/n$, etc.

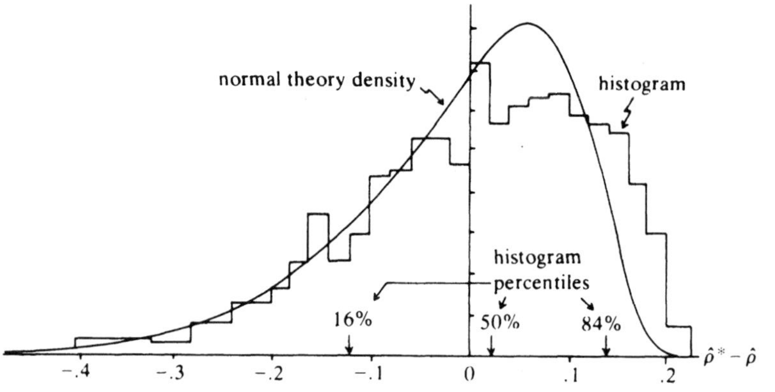

FIG. 5.1. *Histogram of* 1000 *bootstrap replications of* $\hat{\rho}^* - \hat{\rho}$ *for the law school data. The smooth curve is the normal theory density of* $\hat{\rho}^*$, *centered at* $\hat{\rho}$, *when the true correlation is* $\hat{\rho} = .776$.

We could now execute the bootstrap algorithm exactly as before, except starting with $\hat{F}_{\text{NORM}}$ in place of $\hat{F}$ at (5.4). In fact, we don't carry out the Monte Carlo sampling (5.5)–(5.6). Theoretical calculations show that if we did, and if we let $B \to \infty$, $\widehat{\text{SD}}$ as calculated in (5.6) would approximately[2] equal $(1-\hat{\rho}^2)/(n-3)^{1/2}$. Theoretical calculation is impossible outside a narrow family of parametric models, but the bootstrap algorithm (5.4)–(5.6) can always be carried out, given enough raw computing power. The bootstrap is a theory well suited to an age of plentiful computation; see Efron (1979b).

5.3. Smoothed bootstrap. We might wish to attribute some smoothness to F, without going all the way to the normal model (5.7). One way to do this is to use a smoothed estimate of F in place of $\hat{F}$ at step (5.4). In the law school problem, for example, we might use

$$\hat{F}_{.5} = \hat{F} * (.5\,\hat{F}_{\text{NORM}}), \tag{5.8}$$

the convolution of $\hat{F}$ with a version of $\hat{F}_{\text{NORM}}$ scaled down by factor .5. This amounts to smearing out the atoms of $\hat{F}$ into half-sized versions of $\hat{F}_{\text{NORM}}$, each centered at an x_i. Notice that $\hat{F}_{.5}$ has the same correlation coefficient as does $\hat{F}$ and $\hat{F}_{\text{NORM}}$, namely the observed value $\hat{\rho} = .776$. If we were bootstrapping a location parameter, say $\hat{\mu}_y$, instead of $\hat{\rho}$, we would have to rescale $\hat{F}_{.5}$ to have the same covariance matrix as $\hat{F}$. Otherwise our Sd estimate would be biased upward.

Example 5.4. *The correlation coefficient.* Table 5.2, taken from Efron (1980b), is a comparative Monte Carlo study of 15 estimators of standard deviation for

[2] The approximation involved is very much like the approximation of the bootstrap by the jackknife or infinitesimal jackknife, as discussed in Chapter 6.

the correlation coefficient $\hat{\rho}$, and also for $\hat{\phi} = \tanh^{-1} \hat{\rho} = \frac{1}{2} \log (1+\hat{\rho})/(1-\hat{\rho})$. The study comprised 200 trials of $X_1, X_2, \cdots, X_{14} \sim$ bivariate normal with true correlation coefficient $\rho = .5$. Four summary statistics of how the Sd estimates performed in the 200 trials, the average, standard deviation, coefficient of variation and root mean square error (of estimated minus true standard deviation), are presented for each of the 15 estimators, for both $\hat{\rho}$ and $\hat{\phi}$. We will refer back to this table in later chapters as we introduce the estimators of lines 7–14 of the table.

Lines 1 and 2 of Table 5.2 refer to the bootstrap estimate of standard deviation (5.6), with $B = 128$ and 512 respectively. A components of variance analysis revealed that taking $B = \infty$ would not further decrease the root mean square error below .063 for $\widehat{\text{SD}}(\hat{\rho})$ or .061 for $\widehat{\text{SD}}(\hat{\phi})$. Line 3 of the table used the smoothed bootstrap (5.8). Lines 4 and 5 used uniform smoothing: the X_i^* were selected from $\hat{F} * (.5\,\hat{F}_{\text{UNIF}})$, where $\hat{F}_{\text{UNIF}}$ is the uniform distribution over a rhombus selected such that it has the same covariance matrix as $\hat{F}_{\text{NORM}}$. The jackknife results, line 6, are substantially worse, as we have already seen in Table 3.3.

Line 15 gives summary statistics for the parametric bootstrap, i.e., the standard normal theory estimates $\widehat{\text{SD}}(\hat{\rho}) = (1-\hat{\rho}^2)/(n-3)^{1/2}$, $\widehat{\text{SD}}(\hat{\phi}) = [1/(n-3)]^{1/2}$. The ordinary bootstrap performs surprisingly close to the normal theory estimate for $\widehat{\text{SD}}(\hat{\rho})$, so it is not surprising that smoothing doesn't help much here. The motivation behind the $\tanh^{-1}$ transform is to stabilize the variance, that is, to make $\widehat{\text{SD}}$ a constant. In this case smoothing is quite effective. Notice that if the constant .5 in (5.8) were increased toward ∞, the normal smoothed bootstrap would approach the normal theory estimate of Sd, so that the root mean square error would approach zero.

5.4. Bootstrap methods for more general problems. The standard deviation plays no special role in (5.2), or in anything else having to do with the bootstrap. We can consider a perfectly general one-sample problem. Let

$$R(\mathbf{X}, F)$$

be a random variable of interest, where $\mathbf{X} = (X_1, X_2, \cdots, X_n)$ indicates the entire i.i.d. sample $X_1, X_2, \cdots, X_n$. On the basis of having observed $\mathbf{X} = \mathbf{x}$, we wish to estimate some aspect of R's distribution, for example $E_F R$ or $\text{Prob}_F \{R < 2\}$.

The bootstrap algorithm (5.4)–(5.6) is carried out exactly as before, except that at step 2 we calculate

$$R^* = R(\mathbf{X}^*, \hat{F})$$

instead of $\hat{\theta}$, and at step 3 we calculate whichever aspect of R's distribution we are interested in, rather than $\widehat{\text{SD}}$. For instance, if we wish to estimate $E_F R$ we calculate

$$E_* R^* = \frac{1}{B} \sum_{b=1}^{B} R^{*b}, \tag{5.9}$$

TABLE 5.2

A comparison of 15 methods of assigning standard deviation estimates to $\hat{\rho}$ *and* $\hat{\phi} = \tanh^{-1}\hat{\rho}$. *The Monte Carlo experiment consisted of 200 trials of* $X_1, X_2, \cdots, X_{14} \sim$ *bivariate normal, true* $\rho = .5$. *The true standard deviations are* $\mathrm{Sd}(\hat{\rho}) = .218$, $\mathrm{Sd}(\hat{\phi}) = .299$. *Large biases are indicated by asterisks and daggers:* **relative bias* $\geqq .10$, †*relative bias* $\geqq .20$, ‡*relative bias* $\geqq .40$.

Method	$\hat{\rho}$				$\hat{\phi} = \tanh^{-1}\hat{\rho}$			
	AVE	SD	CV	$\sqrt{\text{MSE}}$	AVE	SD	CV	$\sqrt{\text{MSE}}$
1. Bootstrap, $B = 128$	.206	.066	.32	.067	.301	.065	.22	.065
2. Bootstrap, $B = 512$	.206	.063	.31	.064	.301	.062	.21	.062
3. Normal smoothed bootstrap, $B = 128$	.200	.060	.30	.063	.296	.041	.14	.041
4. Uniform smoothed bootstrap, $B = 128$	.205	.061	.30	.062	.298	.058	.19	.058
5. Uniform smoothed bootstrap, $B = 512$	.205	.059	.29	.060	.296	.052	.18	.052
6. Jackknife	.223	.085	.38	.085	.314	.090	.29	.091
7. Infinitesimal jackknife (delta method)	.175†	.058	.33	.072	.244*	.052	.21	.076
8. Half-samples, all 128	.244*	.083	.34	.087	.364†	.099	.27	.118
9. Random HS, $B = 128$	.248*	.079	.32	.085	.368†	.084	.23	.109
10. Balanced HS, 8	.244*	.095	.39	.098	.366†	.111	.30	.129
11. Complementary HS, all 128	.223	.079	.35	.079	.336*	.099	.30	.105
12. Complementary bal. HS, 16	.222	.081	.36	.081	.335*	.100	.30	.106
13. Random subsampling, $B = 128$	.267†	.080	.30	.094	.423‡	.089	.21	.153
14. Random subsampling, range $\widehat{\text{SD}}$	.242	.092	.38	.095	.354*	.077	.27	.111
15. Normal theory	.217	.056	.26	.056	.302	0	0	.003
True value	.218				.299			

while if we are interested in $\text{Prob}_F\{R<2\}$ we calculate

$$\text{Prob}_*\{R^*<2\}=\frac{\#\{R^{*b}<2\}}{B}.$$

In all cases, we are calculating a Monte Carlo approximation to the nonparametric MLE for the quantity of interest, the approximation being that B is finite rather than infinite.

Notation. E_*R^* indicates the expectation of $R^*=R(\mathbf{X}^*,\hat{F})$ under the bootstrap sampling procedure $X_1^*, X_2^*, \cdots, X_n^* \overset{\text{iid}}{\sim} \hat{F}$, $\hat{F}$ fixed as in (5.4); likewise the notation Prob_*, Var_*, Sd_*, etc. Expression (5.9), like (5.6), ignores the fact that B is finite.

5.5. The bootstrap estimate of bias. Suppose we wish to estimate the bias of a functional statistic, $\text{Bias}=E_F\theta(\hat{F})-\theta(F)$ as in (2.3). We can take $R(\mathbf{X},F)=\theta(\hat{F})-\theta(F)$, and use the bootstrap algorithm to estimate $E_FR=\text{Bias}$. In this case

$$R^*=R(\mathbf{X}^*,\hat{F})=\theta(\hat{F}^*)-\theta(\hat{F})=\hat{\theta}^*-\hat{\theta},$$

where $\hat{\theta}^*=\theta(\hat{F}^*)$, $\hat{F}^*$ being the empirical probability distribution of the bootstrap sample: $\hat{F}^*$ puts mass M_i^*/n on x_i, where M_i^* is the number of times x_i appears in the bootstrap sample.

The bootstrap estimate of bias is $\widehat{\text{BIAS}}=E_*R^*$, approximated by

$$\widehat{\text{BIAS}}=\frac{1}{B}\sum_{b=1}^{N}\hat{\theta}^{*b}-\hat{\theta}=\hat{\theta}^{*\cdot}-\hat{\theta}. \tag{5.10}$$

The 1000 bootstrap replications for the law school data yielded $\hat{\rho}^{*\cdot}=.779$, so $\widehat{\text{BIAS}}=.779-.776=.003$, compared to $-.007$ for the jackknife and $-.011$ for normal theory. In the Monte Carlo experiment of Table 5.2, the 200 bootstrap estimates of bias, $B=512$, averaged $-.013$ with standard deviation .022. The jackknife bias estimates averaged $-.018$ with standard deviation .037. The true bias is $-.014$.

We don't need $\hat{\theta}$ to be a functional statistic to apply (5.10), and as a matter of fact we don't need $\hat{\theta}$ to be "the same statistic as θ" in any sense. We could just as well take $\theta(F)=E_FX$ and $\hat{\theta}=$ sample median. To state things as in (5.1), (5.2), write the bias $E_F\hat{\theta}-\theta(F)$ as

$$\text{Bias}=\beta(F,n,\hat{\theta},\theta)=\beta(F),$$

a function of the unknown F, once the sample size n and the forms $\hat{\theta}(\cdot,\cdot,\cdots,\cdot)$ and $\theta(\cdot)$ are fixed. Then the bootstrap estimate is simply

$$\widehat{\text{BIAS}}=\beta(\hat{F}).$$

Chapter 6 indicates the connection between $\widehat{\text{BIAS}}_{\text{BOOT}}$ and $\widehat{\text{BIAS}}_{\text{JACK}}$, the bootstrap and jackknife estimates of Bias. The following theorem is proved in § 6.6.

THEOREM 5.1. *If $\hat{\theta} = \theta(\hat{F})$ is a quadratic functional, then*

$$\widehat{\text{BIAS}}_{\text{BOOT}} = \frac{n-1}{n} \widehat{\text{BIAS}}_{\text{JACK}}.$$

5.6. Finite sample spaces. The rationale for the bootstrap method is particularly evident when the sample space $\mathscr{X}$ is finite, say $\mathscr{X} = \{1, 2, \cdots, L\}$. Then we can express F as $\mathbf{f} = (f_1, f_2, \cdots, f_L)$, where $f_l = \text{Prob}_F\{X = l\}$, and $\hat{F}$ as $\hat{\mathbf{f}} = (\hat{f}_1, \hat{f}_2, \cdots, \hat{f}_L)$, where $\hat{f}_l = \#\{x_i = l\}/n$. The random variable $R(\mathbf{X}, F)$ can be written as

$$R(\mathbf{X}, F) = Q(\hat{\mathbf{f}}, \mathbf{f}), \tag{5.11}$$

some function of $\hat{\mathbf{f}}$ and $\mathbf{f}$, assuming that $R(\mathbf{X}, F)$ is invariant under permutations of the X_i.

The distribution of $\hat{\mathbf{f}}$ given $\mathbf{f}$ is a rescaled multinomial, L categories, n draws and true probability vector $\mathbf{f}$,

$$\hat{\mathbf{f}} \,|\, \mathbf{f} \sim \frac{\text{Mult}_L(n, \mathbf{f})}{n}. \tag{5.12}$$

The bootstrap distribution of $X_1^*, X_2^*, \cdots, X_n^* \overset{\text{iid}}{\sim} \hat{F}$ can be described in terms of $\hat{\mathbf{f}}^* = (\hat{f}_1^*, \hat{f}_2^*, \cdots, \hat{f}_L^*)$, where $\hat{f}_l^* = \#\{X_i^* = l\}/n$. It is the same as (5.12), except that $\hat{\mathbf{f}}$ plays the role of $\mathbf{f}$,

$$\hat{\mathbf{f}}^* \,|\, \hat{\mathbf{f}} \sim \frac{\text{Mult}_L(n, \hat{\mathbf{f}})}{n}. \tag{5.13}$$

The bootstrap method estimates the unobservable distribution of $Q(\hat{\mathbf{f}}, \mathbf{f})$ under (5.12) by the observable distribution of $Q^* = Q(\hat{\mathbf{f}}^*, \hat{\mathbf{f}})$ under (5.13).

Example 5.5. *Binomial probability.* $\mathscr{X} = \{1, 2\}$, $\theta(\mathbf{f}) = f_2 = \text{Prob}_{\mathbf{f}}\{X = 2\}$, $\hat{\theta} = \theta(\hat{\mathbf{f}}) = \hat{f}_2$, and $R(\mathbf{X}, F) = Q(\hat{\mathbf{f}}, \mathbf{f}) = \hat{f}_2 - f_2$, the difference between the observed and theoretical frequency for the second category. From (5.13) we see that Q^* is a standardized binomial,

$$Q^* = \hat{f}_2^* - \hat{f}_2 \underset{*}{\sim} \frac{\text{Bi}\,(n, \hat{f}_2)}{n} - \hat{f}_2$$

with first two moments

$$Q^* \underset{*}{\sim} \left(0, \frac{\hat{f}_1 \hat{f}_2}{n}\right).$$

(The notation "$\underset{*}{\sim}$" indicates the bootstrap distribution.) The implication from the bootstrap theory is that $E_{\mathbf{f}} Q = 0$, i.e., that $\hat{f}_2$ is unbiased for f_2, and $\text{Var}\,\{Q\} = \text{Var}\,\{\hat{f}_2^* - \hat{f}_2\} = (\hat{f}_1 \hat{f}_2)/n$, which of course is the standard binomial estimate.

Asymptotics. As the sample size $n \to \infty$, both $\hat{\mathbf{f}} - \mathbf{f}$ under (5.12) and $\hat{\mathbf{f}}^* - \hat{\mathbf{f}}$ under (5.13) approach the same L-dimensional normal distribution, $\mathcal{N}_L(0, \boldsymbol{\Sigma}_{\mathbf{f}}/n)$, where $\boldsymbol{\Sigma}_{\mathbf{f}}$ has diagonal elements $f_l(1 - f_l)$ and offdiagonal elements $-f_l f_m$. If $Q(\cdot\,, \cdot)$

is a well-behaved function, as described in Efron (1979a, Remark G), then the bootstrap distribution of Q^* is asymptotically the same as the true distribution of Q. This justifies bootstrap inferences, such as estimating $E_F R$ by $E_* R^*$, at least in an asymptotic sense.

What is easy to prove for $\mathscr{X}$ finite is quite difficult for $\mathscr{X}$ more general. Recently Bickel and Freedman (1980) and Singh (1980) have separately demonstrated the asymptotic validity of the bootstrap for $\mathscr{X}$ infinite. They consider statistics like the average U-statistics, t-statistics and quantiles, and show that the bootstrap distribution of R^* converges to the true distribution of R. The convergence is generally quite good, faster than the standard convergence results for the central limit theorem.

5.7. Regression models. So far we have only discussed one-sample situations, where all the random quantities X_i come from the same distribution F. Bootstrap methods apply just as well to many-sample situations, and to a variety of other more complicated data structures. For example, Efron (1980a) presents bootstrap estimates and confidence intervals for censored data. We conclude this chapter with a brief discussion of bootstrap methods for regression models.

A reasonably general regression situation is the following: independent real-valued observations $Y_i = y_i$ are observed, where

$$Y_i = g_i(\beta) + \varepsilon_i, \qquad i = 1, 2, \cdots, n. \tag{5.14}$$

The functions $g_i(\cdot)$ are of known form, usually depending on some observed vector of covariates c_i, while β is a $p \times 1$ vector of unknown parameters. The ε_i are i.i.d. for some distribution F on $\mathscr{R}^1$,

$$\varepsilon_1 \overset{\text{iid}}{\sim} F, \qquad i = 1, 2, \cdots, n, \tag{5.15}$$

where F is assumed to be centered at zero in some sense, perhaps $E_F \varepsilon = 0$ or $\text{Prob}_F\{\varepsilon < 0\} = .5$.

Having observed the $n \times 1$ data vector $\mathbf{Y} = \mathbf{y} = (y_1, y_2, \cdots, y_n)'$, we estimate β by minimizing some measure of distance $D(\mathbf{y}, \boldsymbol{\eta})$ between $\mathbf{y}$ and the vector of predictors $\boldsymbol{\eta}(\beta) = (g_1(\beta), g_2(\beta), \cdots, g_n(\beta))'$,

$$\hat{\beta}: \min_{\beta} D(\mathbf{y}, \boldsymbol{\eta}(\beta)). \tag{5.16}$$

The most common choice of D is $D(\mathbf{y}, \boldsymbol{\eta}) = \sum_{i=1}^{n} (y_i - \eta_i)^2$.

Suppose model (5.14)–(5.16) is too complicated for standard analysis, but we need an assessment of $\hat{\beta}$'s sampling properties. For example, we might have $g_i(\beta) = e^{c_i \beta}$, F of unknown distributional form, and $D(\mathbf{y}, \boldsymbol{\eta}) = \sum |y_i - \eta_i|$. The bootstrap algorithm (5.4)–(5.6) can be modified as follows:

1. Construct $\hat{F}$ putting mass $1/n$ at each observed residual,

$$\hat{F}: \text{mass } \frac{1}{n} \quad \text{at } \hat{\varepsilon}_i = y_i - g_i(\hat{\beta}). \tag{5.17}$$

2. Draw a *bootstrap data set*

(5.18) $$Y_i^* = g_i(\hat{\beta}) + \varepsilon_i^*, \qquad i = 1, 2, \cdots, n,$$

where ε_i^* are i.i.d. from $\hat{F}$, and calculate

(5.19) $$\hat{\beta}^*: \min_{\beta} D(\mathbf{Y}^*, \boldsymbol{\eta}(\beta)).$$

3. Independently repeat step 2 B times, obtaining bootstrap replications $\hat{\beta}^{*1}, \hat{\beta}^{*2}, \cdots, \hat{\beta}^{*B}$.

As an estimate of $\hat{\beta}$'s covariance matrix, for example, we can take

(5.20) $$\widehat{\mathrm{COV}} = \frac{1}{B-1} \sum_{b=1}^{B} (\hat{\beta}^{*b} - \hat{\beta}^{*\cdot})(\hat{\beta}^{*b} - \hat{\beta}^{*\cdot})'.$$

Example 5.6. *Linear regression*. The ordinary linear regression situation is $g_i(\beta) = c_i\beta$, c_i a $1 \times p$ vector of known covariates, and $D(\mathbf{y}, \boldsymbol{\eta}) = \sum (y_i - \eta_i)^2$. Let $\mathbf{C}$ be the $n \times p$ matrix with c_i as the ith row, and $\mathbf{G} = \mathbf{C}'\mathbf{C}$. For convenience assume that the first element of each c_i is 1, and that $\mathbf{G}$ is of full rank p.

In this case we can evaluate (5.20) without recourse to Monte Carlo sampling. Notice that $\hat{F}$ has expected value 0 and variance $\hat{\sigma}^2 = \sum_{i=1}^n \hat{\varepsilon}_i^2/n$, and that $Y_i^* = c_i\hat{\beta} + \varepsilon_i^*$ is a standard linear model written in unusual notation. Standard linear model theory shows that $\hat{\beta}^* = \mathbf{G}^{-1}\mathbf{C}'\mathbf{Y}^*$, and that

(5.21) $$\widehat{\mathrm{COV}} = \hat{\sigma}^2\, \mathbf{G}^{-1}.$$

In other words, the bootstrap gives the standard estimate of covariance in the linear regression case, except for the use of $\sum \hat{\varepsilon}_i^2/n$ rather than $\sum \hat{\varepsilon}_i^2/(n-p)$ to estimate σ^2. This contrasts with the jackknife result (3.13).

The algorithm (5.17)–(5.19) depends on $\hat{F}$ being a reasonable estimate of F, and can give falsely optimistic results if we are fitting highly overparameterized models in hopes of finding a good one. As an example, consider ordinary polynomial regression on the real line. The observed data are of the form (t_1, y_1), $(t_2, y_2), \cdots, (t_n, y_n)$, where t_i is the value of the predictor variable for y_i. If $n = 20$ and we fit an 18th degree polynomial ($g_i(\beta) = c_i\beta$, where $c_i = (1, t_i, t_i^2, \cdots, t_i^{18})$), then $\hat{\sigma}^2 = \sum \hat{\varepsilon}_i^2/20$ is likely to be very small and (5.21) will probably give a foolishly optimistic assessment of $\mathrm{Cov}(\hat{\beta})$. In this case the trouble can be mitigated by using the unbiased estimate of $\sigma^2, \sum \hat{\varepsilon}_i^2$, instead of $\hat{\sigma}^2$, but the general situation is unclear.

As a more cautious alternative to (5.17)–(5.19) we can use the one-sample bootstrap (5.4)–(5.6), with the individual data points being $x_i = (t_i, y_i)$. This method appears to give reasonable answers in model selection situations, as discussed in Chapter 7. On the other hand, it gets us back to results more like (3.13) in the standard linear model.

CHAPTER 6

The Infinitesimal Jackknife, the Delta Method and the Influence Function

In this section we show the connection between the jackknife and the bootstrap, using a simple picture. The picture suggests another version of the jackknife, Jaeckel's (1972) "infinitesimal jackknife". The infinitesimal jackknife turns out to be exactly the same as the ordinary delta method, when the latter applies, and also the same as methods based on the influence function (Hampel (1974)). We begin with a brief discussion of *resampling procedures*, a generic name for all methods which evaluate $\hat{\theta}$ at reweighted versions of the empirical probability distribution $\hat{F}$.

6.1. Resampling procedures. For simplicity, consider a functional statistic $\hat{\theta} = \theta(\hat{F})$, (2.1). The data $x_1, x_2, \cdots, x_n$ are thought of as observed and fixed in what follows. A *resampling vector*[1]

$$\mathbf{P}^* = (P_1^*, P_2^*, \cdots, P_n^*)$$

is any vector on the n-dimensional simplex

$$\mathcal{S}_n = \left\{ \mathbf{P}^* : P_i^* \geqq 0, \sum_{i=1}^{n} P_i^* = 1 \right\}, \tag{6.1}$$

in other words, any probability vector. Corresponding to each $\mathbf{P}^*$ is a reweighted empirical probability distribution $\hat{F}^*$,

$$\hat{F}^*: \text{mass } P_i^* \quad \text{on } x_i, \quad i = 1, 2, \cdots, n, \tag{6.2}$$

and a "resampled" value of $\hat{\theta}$, say $\hat{\theta}^*$,

$$\hat{\theta}^* = \theta(\hat{F}(\mathbf{P}^*)) = \hat{\theta}(\mathbf{P}^*). \tag{6.3}$$

Some of the resampling vectors play special roles in the bootstrap and jackknife theories. In particular,

$$\mathbf{P}^\circ = \left(\frac{1}{n}, \frac{1}{n}, \cdots, \frac{1}{n}\right) \tag{6.4}$$

corresponds to $\hat{F}$ itself, and to the observed value of the statistic $\hat{\theta} = \hat{\theta}(P^\circ)$. The jackknife considers vectors

$$\mathbf{P}_{(i)} = \left(\frac{1}{n-1}, \frac{1}{n-1}, \cdots, 0, \frac{1}{n-1}, \cdots, \frac{1}{n-1}\right) \qquad (0 \text{ in } i\text{th place}), \tag{6.5}$$

[1] It is notationally convenient to consider $\mathbf{P}^*$, as well as some of the other vectors introduced later, as rows rather than columns.

with corresponding values $\hat{\theta}_{(i)}$ of the statistic, $i = 1, 2, \cdots, n$. The bootstrap considers all $\mathbf{P}^*$ vectors of the form $\mathbf{M}^*/n$, $\mathbf{M}^*$ having nonnegative integer coordinates adding to n.

Another way to describe the bootstrap algorithm is to say that the resampling vectors[2] are selected according to a rescaled multinominal distribution,

$$\mathbf{P}^* \underset{*}{\sim} \frac{\text{Mult}_n(n, \mathbf{P}^\circ)}{n}, \tag{6.6}$$

n independent draws on n categories each having probability $1/n$, rescaled by factor $1/n$. The symbol "$\underset{*}{\sim}$" is a reminder that the statistician, not nature, induces the randomness in $\mathbf{P}^*$. Here

$$P_i^* = \frac{\#\{X_j^* = x_i\}}{n},$$

the proportion of the bootstrap sample equal to x_i. The bootstrap standard deviation and bias estimates are simply

$$\widehat{\text{SD}} = \text{Sd}_*\hat{\theta}(\mathbf{P}^*)$$

and

$$\widehat{\text{BIAS}} = E_*\hat{\theta}(\mathbf{P}^*) - \hat{\theta},$$

Sd_* and E_* indicating standard deviation and expectation under (6.6). For future reference note that distribution (6.6) has mean vector and covariance matrix

$$\mathbf{P}^* \underset{*}{\sim} \left(\mathbf{P}^\circ, \frac{\mathbf{I}}{n^2} - \frac{\mathbf{P}^{\circ\prime}\mathbf{P}^\circ}{n}\right), \tag{6.7}$$

where $\mathbf{I}$ is the $n \times n$ identity matrix.

Figure 6.1 shows a schematic representation of the function $\hat{\theta}(\mathbf{P}^*)$ as a curved surface over the simplex $\mathcal{S}_n$. The vertical direction can be taken along the nth coordinate axis since P_n^* is redundant, $P_n^* = 1 - \sum_{i=1}^{n-1} P_1^*$.

Example 6.1. *Quadratic functionals.* A quadratic functional as defined in (4.14), (4.16) is also quadratic as a function of $\mathbf{P}^*$,

$$\hat{\theta}(\mathbf{P}^*) = \hat{\theta}(\mathbf{P}^\circ) + (\mathbf{P}^* - \mathbf{P}^\circ)\mathbf{U} + \tfrac{1}{2}(\mathbf{P}^* - \mathbf{P}^\circ)\mathbf{V}(\mathbf{P}^* - \mathbf{P}^\circ)'. \tag{6.8}$$

The column vector $\mathbf{U}$ and symmetric matrix $\mathbf{V}$ are expressed, after some algebraic manipulation, in terms of $\alpha_i = \alpha^{(\infty)}(x_i)$ and $\beta_{ii'} = \beta_{i'i} = \beta(x_i, x_{i'})$:

$$U_i = \alpha_i - \alpha_{\cdot} + \beta_{i\cdot} - \beta_{\cdot\cdot}, \qquad V_{ii'} = \beta_{ii'} - \beta_{i\cdot} - \beta_{\cdot i'} + \beta_{\cdot\cdot}, \tag{6.9}$$

with the dot indicating averages, $\alpha_{\cdot} = \sum_i \alpha_i/n$, $\beta_{i\cdot} = \sum_{i'} \beta_{ii'}/n$, $\beta_{\cdot\cdot} = \sum_i \beta_{i\cdot}/n$. The expressions in (6.9) satisfy the side conditions

$$\sum_i U_i = 0, \qquad \sum_{i'} V_{ii'} = \sum_i V_{ii'} = 0, \tag{6.10}$$

[2] The symbol $\mathbf{P}^*$ is being used two ways: to indicate a general point in $\mathcal{S}_n$, and also to indicate a random point selected according to (6.6).

which defines the quadratic form (6.8) uniquely. (The possibility of nonuniqueness arises because $\hat{\theta}(\mathbf{P}^*)$ is defined only on $\mathcal{S}_n$, lying in an $(n-1)$-dimensional subspace of R^n.) In the special case $\beta(\cdot,\cdot)=0$, $\hat{\theta}$ is a linear functional statistic,

$$\hat{\theta}(X_1, X_2, \cdots, X_n) = \mu + \frac{1}{n}\sum_{i=1}^{n} \alpha(X_i), \tag{6.11}$$

and $\hat{\theta}(\mathbf{P}^*)$ is a linear function of $\mathbf{P}^*$, $\hat{\theta}(\mathbf{P}^*) = \hat{\theta}(\mathbf{P}^\circ) + (\mathbf{P}^* - \mathbf{P}^\circ)\mathbf{U}$, $U_i = \alpha_i - \alpha_{\cdot}$.

6.2. Relation between the jackknife and bootstrap estimates of standard deviation. There is a unique linear function $\hat{\theta}_{\text{LIN}}(\mathbf{P}^*)$ agreeing with $\hat{\theta}(\mathbf{P}^*)$ at $\mathbf{P}_{(i)}$, $i=1,\cdots,n$,

$$\begin{aligned} \hat{\theta}_{\text{LIN}}(\mathbf{P}^*) &= \hat{\theta}_{(\cdot)} + (\mathbf{P}^* - \mathbf{P}^\circ)\mathbf{U}, \\ U_i &= (n-1)(\hat{\theta}_{(\cdot)} - \hat{\theta}_{(i)}), \qquad i = 1, 2, \cdots, n. \end{aligned} \tag{6.12}$$

THEOREM 6.1. *The jackknife standard deviation estimate for* $\hat{\theta}$, $\widehat{\text{SD}}_{\text{JACK}}(\hat{\theta})$, *is*

$$\widehat{\text{SD}}_{\text{JACK}}(\hat{\theta}) = \sqrt{\frac{n}{n-1}}\ \text{Sd}_*(\hat{\theta}_{\text{LIN}}(\mathbf{P}^*)). \tag{6.13}$$

In other words, $\widehat{\text{SD}}_{\text{JACK}}(\hat{\theta})$ is itself almost a bootstrap Sd estimate, equaling $[n/(n-1)]^{1/2}\ \widehat{\text{SD}}_{\text{BOOT}}(\hat{\theta}_{\text{LIN}})$. The factor $[n/(n-1)]^{1/2}$ makes $[\widehat{\text{SD}}_{\text{JACK}}(\hat{\theta})]^2$ unbiased for $[\text{Sd}(\hat{\theta})]^2$ if $\hat{\theta}$ is a linear functional. As remarked in Chapter 3, we could (but don't) use the same factor to make $[\widehat{\text{SD}}_{\text{BOOT}}(\hat{\theta})]^2$ unbiased in the linear case.

Proof. From (6.12), (6.7) and the fact that $\mathbf{P}^\circ\mathbf{U} = \sum_i U_i/n = 0$,

$$\begin{aligned} \text{Sd}_*\hat{\theta}_{\text{LIN}}(\mathbf{P}^*) &= \left[\frac{1}{n^2}\sum_i U_i^2\right]^{1/2} = \left[\left(\frac{n-1}{n}\right)^2 \sum_i [\hat{\theta}_{(i)} - \hat{\theta}_{(\cdot)}]^2\right]^{1/2} \\ &= \left[\frac{n-1}{n}\right]^{1/2} \widehat{\text{SD}}_{\text{JACK}}(\hat{\theta}). \end{aligned} \tag{6.14}$$ □

Notice that for linear functionals $\hat{\theta}$ we can evaluate $\text{Sd}_*\hat{\theta}(\mathbf{P}^*)$ directly from (6.7), without resorting to Monte Carlo calculations. The jackknife requires less computation than the bootstrap because it approximates any $\hat{\theta}$ with a linear functional.

Figure 6.1 is misleading in one important sense. Under (6.6), $\|\mathbf{P}^* - \mathbf{P}\| = O_p(1/\sqrt{n})$, while $\|\mathbf{P}_{(i)} - \mathbf{P}^\circ\| = O(1/n)$. The bootstrap resampling vectors tend to be much further away from the central value $\mathbf{P}^\circ$ than are the jackknife resampling vectors $\mathbf{P}_{(i)}$. This is what causes trouble for markedly nonlinear statistics like the median. The grouped jackknife, § 2.2, resamples at a distance $(1/n)[hg/(g-1)]^{1/2}$ from $\mathbf{P}^\circ$. Taking the group size $h = n^{1/2}$ gives distance $O(1/\sqrt{n})$, as with the bootstrap, and gives an asymptotically correct Sd estimate for the sample median.

6.3. Jaeckel's infinitesimal jackknife. Figure 6.1 suggests another estimate of standard deviation: instead of approximating $\hat{\theta}(\mathbf{P}^*)$ by $\hat{\theta}_{\text{LIN}}(\mathbf{P}^*)$, why not use

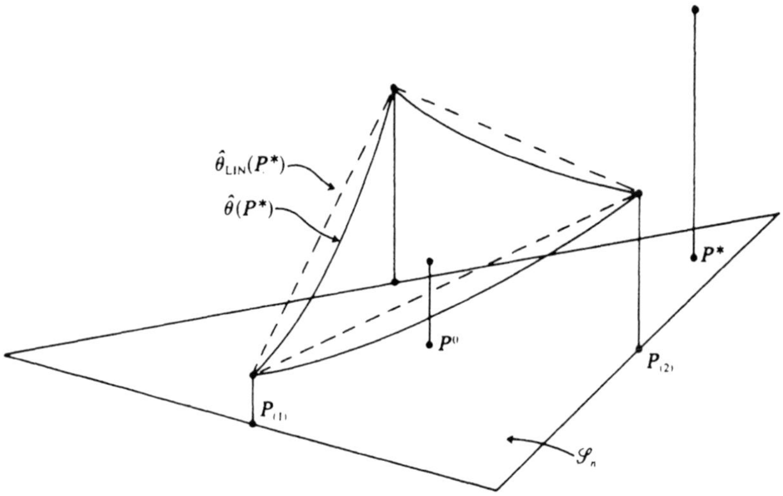

FIG. 6.1. *A schematic representation of* $\hat{\theta}(\mathbf{P}^*)$ *as a function on* $\mathscr{S}_n$. *The curved surface* $\hat{\theta}(\cdot)$ *is approximated by the linear function* $\hat{\theta}_{LIN}(\cdot)$. *The bootstrap standard deviation estimate is* $\mathrm{Sd}_*\hat{\theta}(\mathbf{P}^*)$, *the jackknife* Sd *estimate is* $[n/(n-1)]^{1/2}\,\mathrm{Sd}_*\hat{\theta}_{LIN}(\mathbf{P}^*)$, *where* Sd_* *indicates standard deviation under the multinomial distribution* (6.6).

$\hat{\theta}_{TAN}(\mathbf{P}^*)$, the hyperplane tangent to the surface $\hat{\theta}(\mathbf{P}^*)$ at the point $\mathbf{P}^* = \mathbf{P}^\circ$? The corresponding estimate of standard deviation is then

$$\widehat{\mathrm{SD}} = \mathrm{Sd}_*\hat{\theta}_{TAN}(\mathbf{P}^*), \tag{6.15}$$

Sd_* indicating standard deviation under distribution (6.6) as before. Formula (6.15) is Jaeckel's (1972) *infinitesimal jackknife* estimate of standard deviation. In other words, $\widehat{\mathrm{SD}}_{IJ}(\hat{\theta}) = \widehat{\mathrm{SD}}_{BOOT}(\hat{\theta}_{TAN})$.

The function $\hat{\theta}_{TAN}(\cdot)$ equals

$$\hat{\theta}_{TAN}(\mathbf{P}^*) = \hat{\theta}(\mathbf{P}^\circ) + (\mathbf{P}^* - \mathbf{P}^\circ)\mathbf{U},$$
$$U_i = \lim_{\varepsilon \to 0} \frac{\hat{\theta}(\mathbf{P}^\circ + \varepsilon(\boldsymbol{\delta}_i - \mathbf{P}^\circ)) - \hat{\theta}(\mathbf{P}^\circ)}{\varepsilon}, \qquad i = 1, 2, \cdots, n, \tag{6.16}$$

$\boldsymbol{\delta}_i$ being the ith coordinate vector. The U_i are essentially directional derivatives. Suppose we extend the definition of $\hat{\theta}(\mathbf{P}^*)$ to values of $\mathbf{P}^*$ outside $\mathscr{S}_n$ in any reasonable way, for example by the homogeneous extension $\hat{\theta}(\mathbf{P}^*) = \hat{\theta}(\mathbf{Q}^*)$, $\mathbf{Q}^* = \mathbf{P}^*/\sum_{i=1}^n P_i^*$. (If $\mathbf{P}^*$ has nonnegative coordinates summing to any positive value then $\mathbf{Q}^* \in \mathscr{S}_n$; the homogeneous extension assigns $\hat{\theta}(\mathbf{P}^*)$ the same values along any ray from the origin.) Let $\mathbf{D}$, a column vector, be the gradient vector of $\hat{\theta}(\mathbf{P}^*)$ evaluated at $\mathbf{P}^* = \mathbf{P}^\circ$,

$$D_i = \frac{\partial}{\partial P_i^*} \hat{\theta}(P^*)\Big|_{\mathbf{P}^* = \mathbf{P}^\circ}.$$

This definition makes sense because $\hat{\theta}(\mathbf{P}^*)$ is now defined in an open neighborhood of $\mathbf{P}^\circ$. Then $U_i = (\boldsymbol{\delta}_i - \mathbf{P}^\circ)\mathbf{D}$, and we see that

$$\mathbf{P}^\circ \mathbf{U} = \frac{1}{n} \sum_{i=1}^{n} U_i = 0. \tag{6.17}$$

Therefore, applying (6.7) to (6.15), (6.16) shows that the infinitesimal jackknife estimate of standard deviation is

$$\widehat{\mathrm{SD}}_{\mathrm{IJ}}(\hat{\theta}) = \left[\frac{1}{n^2} \sum_{i=1}^{n} U_i^2\right]^{1/2}. \tag{6.18}$$

The infinitesimal jackknife resamples $\hat{\theta}$ at $\mathbf{P}^*$ values infinitesimally close to $\mathbf{P}^\circ$, rather than $O(1/n)$ away as with the ordinary jackknife—hence the name "infinitesimal". Looking at the last equation in (6.16), take $\varepsilon = -1/(n-1)$ instead of letting $\varepsilon \to 0$. Then $U_i = [\hat{\theta}(\mathbf{P}^\circ + \varepsilon(\boldsymbol{\delta}_i - \mathbf{P}^\circ)) - \hat{\theta}(\mathbf{P}^\circ)]/\varepsilon = (n-1)(\hat{\theta} - \hat{\theta}_{(i)})$, which is similar to definition (6.12) (compare (6.14) with (6.18)).

We can use other values of ε to define other jackknives. For example, taking $\varepsilon = 1/(n+1)$ in (6.16) makes $U_i = (n+1)(\hat{\theta}_{[i]} - \hat{\theta})$, where $\hat{\theta}_{[i]} = \hat{\theta}(x_1, x_2, \cdots, x_i, x_i, x_{i+1}, \cdots, x_n)$, i.e., the value of $\hat{\theta}$ when x_i is repeated rather than removed from the data set. The corresponding standard deviation estimate $[\sum (U_i - U_\cdot)^2/n^2]^{1/2}$, which might be called the "positive jackknife", is the bootstrap Sd of the linear function having value $\hat{\theta}_{[i]}$ at $P_{[i]} = (1/n, 1/n, \cdots, 2/n, \cdots, 1/n)$, $2/n$ in the ith place, $i = 1, 2, \cdots, n$. The positive jackknife was applied to $\hat{\theta}$ the correlation coefficient in Hinkley (1978), and produced estimates of standard deviation with extreme downward biases.

Getting back to the infinitesimal jackknife, consider a linear functional statistic $\hat{\theta} = \mu + \sum \alpha(X_i)/n$ having representation (6.3), $\hat{\theta}(\mathbf{P}^*) = \mu + \sum P_i^* \alpha_i$. Then U_i as defined in (6.16) equals $\alpha_i - \alpha_\cdot$, $\alpha_\cdot = \sum \alpha_i/n$, so

$$\widehat{\mathrm{SD}}_{\mathrm{IJ}} = \left[\frac{1}{n^2} \sum_{i=1}^{n} (\alpha_i - \alpha_\cdot)^2\right]^{1/2}.$$

Definition (6.15) does not include the bias correction factor $\sqrt{n/(n-1)}$, so for linear functionals $E_F[\widehat{\mathrm{SD}}_{\mathrm{IJ}}]^2 = ((n-1)/n)[\mathrm{Sd}]^2$. Multiplying $\widehat{\mathrm{SD}}_{\mathrm{IJ}}$ by $\sqrt{14/13} = 1.038$ helps correct the severe downward bias of the infinitesimal jackknife evident in line 7 of Table 5.2, but not by much.

The directional derivatives U_i in (6.16) can be calculated explicitly for many common statistics. For example, if the statistic is the sample correlation coefficient $\hat{\rho}$, then

$$U_i = -\frac{1}{2}\hat{\rho} \cdot \left[\left(\frac{y_i - \hat{\mu}_y}{\sqrt{\hat{\mu}_{yy}}}\right)^2 + \left(\frac{z_i - \hat{\mu}_z}{\sqrt{\hat{\mu}_{zz}}}\right)^2\right] + \left(\frac{y_i - \hat{\mu}_y}{\sqrt{\hat{\mu}_{yy}}}\right)\left(\frac{z_i - \hat{\mu}_z}{\sqrt{\hat{\mu}_{zz}}}\right), \tag{6.19}$$

where $x_i = (y_i, z_i)$, $\hat{\mu}_y = \sum y_i/n$, $\hat{\mu}_{yy} = \sum (y_i - \hat{\mu}_y)^2/n$, etc. In fact, it is usually more convenient to evaluate the U_i numerically: simply substitute a small value of ε into definition (6.16). The value of $\varepsilon = .001$ was used in Table 5.2.

6.4. Influence function estimates of standard deviation. The influence function expansion (4.13), $\theta(\hat{F}) = \theta(F) + \sum \text{IF}(X_i)/n + O_p(1/n)$, approximates an arbitrary functional statistic $\theta(\hat{F})$ by a constant $\theta(F)$ plus an average of i.i.d. random variables $\sum \text{IF}(X_i)/n$. This immediately suggests the standard deviation approximation

$$\text{Sd}(\hat{\theta}) \doteq \left[\text{Var}_F \frac{\text{IF}(X)}{n}\right]^{1/2}, \tag{6.20}$$

where, since $E_F \text{IF}(X) = 0$ (basically the same result as (6.17), see Hampel (1974, § 3.5)),

$$\text{Var}_F \text{IF}(X) = \int_{\mathscr{X}} \text{IF}^2(x)\, dF(x).$$

In order to use (6.20) as an Sd estimator, we have to estimate $\text{Var}_F\ \text{IF}(X)$. The definition $\text{IF}(x) = \lim_{\varepsilon \to 0} [\theta((1-\varepsilon)F + \varepsilon\delta_x) - \theta(F)]/\varepsilon$ is obviously related to the definition of U_i in (6.16). As a matter of fact, U_i is the influence function of $\theta(F)$ for $F = \hat{F}$, evaluated at $x = x_i$. Mallows (1974) aptly calls U_i the *empirical influence function*, denoted $\widehat{\text{IF}}(x_i)$. The obvious estimate of $\text{Var}_F\ \text{IF}(X)$ is $\int_{\mathscr{X}} \widehat{\text{IF}}^2(x)\, d\hat{F}(x) = \sum U_i^2/n$. Plugging this into (6.20) gives the Sd estimate $[\sum U_i^2/n^2]^{1/2}$, which is exactly the infinitesimal jackknife estimate (6.18).

The ordinary jackknife and positive jackknife can also be thought of as estimates of (6.20). They use the influence function estimates

$$\widehat{\text{IF}}(x_i) = \begin{cases} (n-1)(\hat{\theta}_{(\cdot)} - \hat{\theta}_{(i)}) & \text{(ordinary jackknife)}, \\ (n+1)(\hat{\theta}_{[i]} - \hat{\theta}_{[\cdot]}) & \text{(positive jackknife)}. \end{cases}$$

All these methods give asymptotically correct results if $\theta(\cdot)$ is sufficiently smooth, but perform quite differently in small samples, as illustrated by Table 5.2. In the author's experience, the ordinary jackknife is the only jackknife which can be trusted not to give badly underbiased estimates of standard deviation. (This is the import of Theorem 4.1.) If one isn't going to use the bootstrap, because of computational costs, the ordinary jackknife seems to be the method of choice.

6.5. The delta method. Many statistics are of the form

$$\hat{\theta}(X_1, X_2, \cdots, X_n) = t(\bar{Q}_1, \bar{Q}_2, \cdots, \bar{Q}_A), \tag{6.21}$$

where $t(\cdot, \cdot, \cdots, \cdot)$ is a known function and each $\bar{Q}_a$ is an observed average

$$\bar{Q}_a = \frac{1}{n} \sum_{i=1}^{n} Q_a(X_i).$$

For example, the correlation coefficient $\hat{\rho}$ equals

$$t(\bar{Q}_1, \bar{Q}_2, \bar{Q}_3, \bar{Q}_4, \bar{Q}_5) = \frac{\bar{Q}_4 - \bar{Q}_1\bar{Q}_2}{[\bar{Q}_3 - \bar{Q}_1^2]^{1/2}[\bar{Q}_5 - \bar{Q}_2^2]^{1/2}},$$

with $Q_1(X) = Q_1((Y, Z)) = Y$, $Q_2 = Z$, $Q_3 = Y^2$, $Q_4 = YZ$, $Q_5 = Z^2$.

Suppose that the vector $\mathbf{Q}(X) = (Q_1(X), Q_2(X), \cdots, Q_A(X))$, corresponding to one observation of $X \sim F$, has a mean vector $\boldsymbol{\alpha}_F$ and covariance matrix $\boldsymbol{\beta}_F$, and let $\boldsymbol{\nabla}_F$ be the gradient vector with ath component $\partial t/\partial Q_a|_{\mathbf{Q}=\boldsymbol{\alpha}_F}$. Expanding $\hat{\theta} = t(\bar{\mathbf{Q}})$ in a first order Taylor series about $\boldsymbol{\alpha}_F$ gives the approximation

$$\text{(6.22)} \qquad \text{Sd}(\hat{\theta}) \doteq \left[\frac{\boldsymbol{\nabla}_F \boldsymbol{\beta}_F \boldsymbol{\nabla}'_F}{n}\right]^{1/2}.$$

In the case of the correlation coefficient somewhat tedious calculations show that (6.22) becomes

$$\text{(6.23)} \quad \text{Sd}(\hat{\rho}) \doteq \left\{\frac{\rho^2}{4n}\left[\frac{\mu_{40}}{\mu_{20}^2} + \frac{\mu_{04}}{\mu_{02}^2} + \frac{2\mu_{22}}{\mu_{20}\mu_{02}} + \frac{4\mu_{22}}{\mu_{11}^2} - \frac{4\mu_{31}}{\mu_{11}\mu_{20}} - \frac{4\mu_{13}}{\mu_{11}\mu_{02}}\right]\right\}^{1/2},$$

where, denoting $X = (Y, Z)$, $\mu_{gh} = E_F[Y - E_F Y]^g[Z - E_F Z]^h$.

Substituting $\hat{F}$ for F in (6.22) gives the *nonparametric delta method* estimate of standard deviation,

$$\text{(6.24)} \qquad \widehat{\text{SD}} = \left[\frac{\boldsymbol{\nabla}_{\hat{F}} \boldsymbol{\beta}_{\hat{F}} \boldsymbol{\nabla}'_{\hat{F}}}{n}\right]^{1/2}.$$

For example (6.23) would be estimated by the same expression with $\hat{\rho}$ replacing ρ and the sample moments $\hat{\mu}_{gh}$ replacing the $\hat{\mu}_{gh}$.

THEOREM 6.2. *For any statistic of form* (6.21), *the nonparametric delta method and the infinitesimal jackknife give identical estimates of standard deviation.*

Proof. For statistics $\hat{\theta}$ of form (6.21), the directional derivatives U_i in (6.16) are

$$U_i = \boldsymbol{\nabla}_{\hat{F}} \cdot [\mathbf{Q}(x_i) - \bar{\mathbf{Q}}],$$

since

$$\hat{\theta}(\mathbf{P}^\circ + \varepsilon(\boldsymbol{\delta}_i - \mathbf{P}^\circ)) = t(\bar{\mathbf{Q}} + \varepsilon(\mathbf{Q}(x_i) - \bar{\mathbf{Q}})) \doteq t(\bar{\mathbf{Q}}) + \varepsilon \boldsymbol{\nabla}_{\hat{F}} \cdot [\mathbf{Q}(x_i) - \bar{\mathbf{Q}}].$$

Therefore (6.18) gives

$$\widehat{\text{SD}}_{\text{IJ}} = \left[\boldsymbol{\nabla}_{\hat{F}} \frac{1}{n} \sum [\mathbf{Q}(x_i) - \bar{\mathbf{Q}}]'[\mathbf{Q}(x_i) - \bar{\mathbf{Q}}] \boldsymbol{\nabla}'_{\hat{F}} \Big/ n\right]^{1/2} = \left[\frac{\boldsymbol{\nabla}_{\hat{F}} \boldsymbol{\beta}_{\hat{F}} \boldsymbol{\nabla}'_{\hat{F}}}{n}\right]^{1/2},$$

agreeing with the delta method estimate (6.24). Here we have used the fact that

$$\frac{1}{n} \sum [Q(x_i) - \bar{Q}]'[Q(x_i) - \bar{Q}] = \text{Cov}_{F=\hat{F}} \mathbf{Q},$$

the covariance of $\mathbf{Q}$ under $\hat{F}$, and so must equal $\boldsymbol{\beta}_{\hat{F}}$. Likewise

$$\left(\cdots, \frac{\partial t}{\partial Q_a}, \cdots\right)\Big|_{\mathbf{Q}=\bar{\mathbf{Q}}} = \left(\cdots, \frac{\partial t}{\partial Q_a}, \cdots\right)\Big|_{\mathbf{Q}=\boldsymbol{\alpha}_{\hat{F}}} = \boldsymbol{\nabla}_{\hat{F}}. \qquad \square$$

By comparing (6.18), (6.19) with (6.23), (6.24) the reader can see how Theorem 6.2 works for the correlation coefficient $\hat{\rho}$. Both (6.19) and (6.23) are difficult calculations, fraught with the possibility of error, and it is nice to know that they can be circumvented by numerical methods, as we commented at the end of § 6.3.

The infinitesimal jackknife works by perturbing the weights $1/n$ which define $\hat{F}$, keeping the x_i fixed, and seeing how $\hat{\theta}$ varies. The delta method perturbs the x_i (which only affect $\hat{\theta}$ through the $\bar{Q}_a$ in form (6.21)), keeping the weights constant. It is reassuring to see that the results are identical. In this sense there is only one nonparametric delta method.

So far we have discussed the delta method in a nonparametric framework. Suppose, on the other hand, that F is known to belong to a parametric family, say $\mathscr{F}$.

$$\mathscr{F} = \{F_\theta : \theta \in \Theta\}, \tag{6.25}$$

Θ a subset of R^p. Write $\boldsymbol{\alpha}_\theta$, $\boldsymbol{\beta}_\theta$ and $\boldsymbol{\nabla}_\theta$ in place of $\boldsymbol{\alpha}_{F_\theta}$, $\boldsymbol{\beta}_{F_\theta}$, $\boldsymbol{\nabla}_{F_\theta}$. The *parametric delta method* estimate of standard deviation for a statistic $t(\bar{\mathbf{Q}})$ is

$$\widehat{\text{SD}} = \left[\frac{\boldsymbol{\nabla}_{\hat{\theta}}\boldsymbol{\beta}_{\hat{\theta}}\boldsymbol{\nabla}'_{\hat{\theta}}}{n}\right]^{1/2}, \tag{6.26}$$

and is closely related to the parametric bootstrap of § 5.2. Here $\hat{\theta}$ is the MLE of θ. Without pursuing the details, we mention that (6.26) can be applied to the case where $t(\bar{\mathbf{Q}})$ is the MLE of some parameter, and results in the familiar Fisher information bound for the standard deviation of a maximum likelihood estimator. Jaeckel (1972) discussed this point.

6.6. Estimates of bias. Figure 6.1 also helps relate the jackknife and bootstrap estimates of bias.

THEOREM 6.3. *Let* $\hat{\theta}_{\text{QUAD}}(\mathbf{P}^*)$ *be any quadratic function* $\mathbf{a} + (\mathbf{P}^* - \mathbf{P}^\circ)\mathbf{b} + \frac{1}{2}(\mathbf{P}^* - \mathbf{P}^\circ)\mathbf{c}(\mathbf{P}^* - \mathbf{P}^\circ)'$ ($\mathbf{a}$ *a constant,* $\mathbf{b}$ *a column vector,* $\mathbf{c}$ *a symmetric matrix*) *satisfying*

$$\hat{\theta}_{\text{QUAD}}(\mathbf{P}^\circ) = \hat{\theta}(\mathbf{P}^\circ) \quad \text{and} \quad \hat{\theta}_{\text{QUAD}}(\mathbf{P}_{(i)}) = \hat{\theta}(\mathbf{P}_{(i)}), \qquad i = 1, 2, \cdots, n. \tag{6.27}$$

Then

$$\widehat{\text{BIAS}}_{\text{JACK}}(\hat{\theta}) = \frac{n}{n-1}[E_* \hat{Q}_{\text{QUAD}}(\mathbf{P}^*) - \hat{\theta}_{\text{QUAD}}(\mathbf{P}^\circ)]. \tag{6.28}$$

In the other words, the jackknife estimate of bias equals $n/(n-1)$ times the bootstrap estimate of bias for any quadratic function agreeing with $\hat{\theta}$ at $\mathbf{P}^\circ$ *and* $\mathbf{P}_{(1)}, \mathbf{P}_{(2)}, \cdots, \mathbf{P}_{(n)}$.

Proof. We can always rewrite a quadratic function $\hat{\theta}_{\text{QUAD}}(\mathbf{P}^*) = a + (\mathbf{P}^* - \mathbf{P})\mathbf{b} + \frac{1}{2}(\mathbf{P}^* - \mathbf{P}^\circ)\mathbf{c}(\mathbf{P}^* - \mathbf{P}^\circ)'$ so that

$$\mathbf{P}^\circ\mathbf{b} = 0, \qquad \mathbf{P}^\circ\mathbf{c} = 0. \tag{6.29}$$

(If (6.29) is not satisfied, replace $\mathbf{b}$ with $\mathbf{b} - b_{.}\mathbf{1}$, where $b_{.} = \mathbf{P}^\circ\mathbf{b}$ and $\mathbf{1} = (1, 1, \cdots, 1)'$; replace $\mathbf{c}$ with $\mathbf{c} - \mathbf{c}_{.}\mathbf{1}' - \mathbf{1}\mathbf{c}_{.}' + \mathbf{1}c_{..}\mathbf{1}'$, where $\mathbf{c}_{.} = \mathbf{c}\mathbf{P}^{\circ\prime}$, $c_{..} = \mathbf{P}^\circ\mathbf{c}\mathbf{P}^{\circ\prime}$.)

From (6.7) we compute

$$E_*\hat{\theta}_{\text{QUAD}}(\mathbf{P}^*)-\hat{\theta}_{\text{QUAD}}(\mathbf{P}^\circ)=\frac{1}{2}\operatorname{tr}\mathbf{c}\left(\frac{\mathbf{I}}{n^2}-\frac{\mathbf{P}^{\circ\prime}\mathbf{P}^\circ}{n}\right)=\frac{1}{2}\operatorname{tr}\frac{\mathbf{c}}{n^2}=\frac{1}{2n^2}\sum c_{ii} \tag{6.30}$$

By (6.27),

$$\hat{\theta}(\mathbf{P}_{(i)})-\hat{\theta}(\mathbf{P}^\circ)=\hat{\theta}_{\text{QUAD}}(\mathbf{P}_{(i)})-\hat{\theta}_{\text{QUAD}}(\mathbf{P}^\circ)=-\frac{b_i}{n-1}+\frac{1}{2}\frac{c_{ii}}{(n-1)^2}, \tag{6.31}$$

the last result following from $\mathbf{P}_{(i)}-\mathbf{P}^\circ=(\mathbf{P}^\circ-\boldsymbol{\delta}_i)/(n-1)$ and (6.29). Averaging (6.31) over i, and using $b_.=\mathbf{P}^\circ\mathbf{b}=0$ again, gives

$$\widehat{\text{BIAS}}_{\text{JACK}}(\hat{\theta})=\frac{1}{2n(n-1)}\sum c_{ii}. \tag{6.32}$$

Comparing (6.32) with (6.30) verifies (6.28). □

We have seen, in (6.8), (6.9), that a quadractic functional statistic $\theta(\hat{F})$ is also a quadratic function $\hat{\theta}(\mathbf{P}^*)$. In this case we can take $\hat{\theta}_{\text{QUAD}}=\hat{\theta}$, so (6.28) becomes

$$\widehat{\text{BIAS}}_{\text{JACK}}(\hat{\theta})=\frac{n}{n-1}\widehat{\text{BIAS}}_{\text{BOOT}}(\hat{\theta})$$

for $\hat{\theta}$ quadratic, which is Theorem 5.1. The factor $n/(n-1)$ makes $\widehat{\text{BIAS}}_{\text{JACK}}$ unbiased for the true bias of a quadratic functional, Theorem 2.1. Notice the similarity of this result to Theorem 6.1.

The infinitesimal jackknife and nonparametric delta method give identical estimates of bias, just as in Theorem 6.2. These estimates are $\sum V_{ii}/2n^2$ and $\frac{1}{2}\operatorname{tr}\boldsymbol{\beta}_{\hat{F}}\boldsymbol{\nabla}^2_{\hat{F}}$ respectively, where $\boldsymbol{\nabla}^2_{\hat{F}}$ is the $A\times A$ matrix with a,bth element

$$\left.\frac{\partial^2 t}{\partial Q_a\,\partial Q_b}\right|_{\mathbf{Q}=\boldsymbol{\alpha}_{\hat{F}}}\quad\text{and}\quad V_{ii}=\left.\frac{d^2\hat{\theta}(\mathbf{P}^\circ+\varepsilon(\delta_i-\mathbf{P}^\circ))}{d\varepsilon^2}\right|_{\varepsilon=0}.$$

A proof is given in Gray, Schucany and Watkins (1975). If we can expand $\hat{\theta}(\mathbf{P}^*)$ in a Taylor series about $\mathbf{P}^\circ$,

$$\hat{\theta}(\mathbf{P}^*)=\hat{\theta}(\mathbf{P}^\circ)+(\mathbf{P}^*-\mathbf{P}^\circ)\mathbf{U}+\tfrac{1}{2}(\mathbf{P}^*-\mathbf{P}^\circ)\mathbf{V}(\mathbf{P}^*-\mathbf{P}^\circ)'+\cdots,$$

then stopping the series after the quadratic term gives a quadratic approximation $\hat{\theta}_Q(\mathbf{P}^*)$ to $\hat{\theta}(\mathbf{P}^*)$. The infinitesimal jackknife estimate of bias equals the bootstrap estimate of bias for $\hat{\theta}_Q$,

$$\widehat{\text{BIAS}}_{\text{IJ}}(\hat{\theta})=\widehat{\text{BIAS}}_{\text{BOOT}}(\hat{\theta}_Q),$$

just as after (6.15).

6.7. More general random variables. So far we have considered functional statistics $\theta(\hat{F})$. More generally we might be interested in a random quantity

$$R(\hat{F},F), \tag{6.33}$$

for example the Kolmogorov–Smirnov test statistic on $\mathscr{X} = \mathscr{R}^1$,

$$\sup_x \left| \frac{\#\{X_i \leqq x\}}{n} - \text{Prob}_F\{X \leqq x\} \right|.$$

The resampled quantity R^* corresponding to R is

$$R^* = R(\hat{F}^*, \hat{F}) = R(P^*). \tag{6.34}$$

Here $\hat{F}^*$ is the reweighted empirical distribution (6.2). The shorthand notation $R(\mathbf{P}^*)$ tacitly assumes that $x_1, x_2, \cdots, x_n$ are fixed at their observed values.

The curved surface in Fig. 6.1 is now $R(\mathbf{P}^*)$ rather than $\hat{\theta}(\mathbf{P}^*)$. The bootstrap estimate of any quantity of interest, such as $E_F R(\hat{F}, F)$ or $\text{Prob}_F\{R(\hat{F}, F) > 2/\sqrt{n}\}$, is the corresponding quantity computed under (6.6), e.g. $E_* R(\mathbf{P}^*)$ or $\text{Prob}_*\{R(\mathbf{P}^*) > 2/\sqrt{n}\}$. Jackknife approximations can be used as before to reduce the bootstrap computations:

$$\text{Sd}_*(R(\mathbf{P}^*)) \doteq \left[\frac{n-1}{n} \sum_{i=1}^{n} [R_{(i)} - R_{(\cdot)}]^2\right]^{1/2}, \tag{6.35}$$

and

$$E_* R(\mathbf{P}^*) - R(\mathbf{P}^\circ) \doteq (n-1)(R_{(\cdot)} - R(\mathbf{P}^\circ)), \tag{6.36}$$

$R_{(i)} = R(\mathbf{P}_{(i)})$, $R_{(\cdot)} = \sum R_{(i)}/n$.

The justification of (6.35), (6.36) is the same as that given for Theorems 6.1 and 6.3: If $R_{\text{LIN}}(\mathbf{P}^*)$ is the linear function of $\mathbf{P}^*$ agreeing with $R(\mathbf{P}^*)$ for $\mathbf{P}^* = \mathbf{P}_{(i)}$, $i = 1, 2, \cdots, n$, then the right side of (6.35) equals $[n/(n-1)]^{1/2}\,\text{Sd}_*(R_{\text{LIN}}(\mathbf{P}^*))$. Likewise, the right side of (6.36) equals $(n/(n-1))[E_* R_{\text{QUAD}}(\mathbf{P}^*) - R(\mathbf{P}^\circ)]$, where $R_{\text{QUAD}}(\mathbf{P}^*)$ is any quadratic function agreeing with $R(\mathbf{P}^*)$ at $\mathbf{P}^\circ, \mathbf{P}_{(1)}, \mathbf{P}_{(2)}, \cdots, \mathbf{P}_{(n)}$. We could use the more direct approximation

$$\text{Sd}_*(R(\mathbf{P}^*)) \doteq \text{Sd}_*(R_{\text{LIN}}(\mathbf{P}^*)) = \left[\left(\frac{n-1}{n}\right)^2 \sum [R_{(i)} - R_{(\cdot)}]^2\right]^{1/2},$$

instead of (6.35), and make the corresponding change in (6.36), but we won't do so in what follows.

The infinitesimal jackknife also yields approximations to the bootstrap standard deviation and expectation.

$$\text{Sd}_*(R(\mathbf{P}^*)) \doteq \left[\frac{1}{n^2} \sum U_i^2\right]^{1/2}, \qquad E_* R(\mathbf{P}^*) - R(\mathbf{P}^\circ) = \frac{1}{2n} \sum V_{ii}. \tag{6.37}$$

Here U_i and V_{ii} are defined as in § 6.6, with $R(\mathbf{P}^\circ + \varepsilon(\delta_i - \mathbf{P}^\circ))$ replacing $\hat{\theta}(\mathbf{P}^\circ + \varepsilon(\delta_i - \mathbf{P}^\circ))$.

What happens if we consider variables not of the functional form $R(\hat{F}, F)$? As an example, consider $R = a_n + b_n \hat{\theta} - \theta$, where $\hat{\theta} = \theta(\hat{F})$ is a functional statistic, $\theta = \theta(F)$, and a_n and b_n are constants, $\lim a_n = 0$, $\lim b_n = 1$. The bootstrap estimate of bias (for $a_n + b_n\hat{\theta}$ as an estimate of θ), as discussed in § 5.5 is

$$E_*R^* = a_n + b_n E_*\hat{\theta}^* - \hat{\theta} = a_n + (b_n - 1)\hat{\theta} + b_n(E_*\hat{\theta}^* - \hat{\theta})$$
$$= a_n + (b_n - 1)\hat{\theta} + b_n \widehat{\mathrm{BIAS}}_{\mathrm{BOOT}}(\hat{\theta}),$$

compared with the true bias

$$a_n + (b_n - 1)\theta + b_n(E_F\hat{\theta} - \theta) = a_n + (b_n - 1)\theta + b_n \mathrm{Bias}(\hat{\theta}). \tag{6.38}$$

Define

$$A_n = na_n, \qquad B_n = nb_n.$$

Then the jackknife estimate of bias reduces to

$$a_n + (b_n - 1)\hat{\theta} + b_{n-1}\widehat{\mathrm{BIAS}}_{\mathrm{JACK}}(\hat{\theta}) + [(A_{n-1} - A_n) + (B_{n-1} - B_n + 1)\hat{\theta}]. \tag{6.39}$$

The situation is quite delicate: if $a_n = c/n$ and $b_n = 1 - d/n$, then $A_{n-1} - A_n = 0$, $B_{n-1} - B_n = 0$, and (6.39) agrees nicely with (6.38). On the other hand, if $a_n = (-1)^n a/n$, $b_n = 1 - d/n$, expansion (6.39) can completely disagree with (6.38). The jackknife estimate of bias is not recommended for statistics other than functional form.

CHAPTER 7

Cross Validation, the Jackknife and the Bootstrap

Cross validation is an old idea whose time seems to have come again with the advent of modern computers. The original method goes as follows. We are interested in fitting a regression model to a set of data, but are not certain of the model's form, e.g., which predictor variables to include, whether or not to make a logarithmic transform on the response variable, which interactions to include if any, etc. The data set is randomly divided into two halves, and the first half used for model fitting. Anything goes during this phase of the procedure, including hunches, preliminary testing, looking for patterns, trying large numbers of different models and eliminating "outliers."

The second phase is the cross validation: the regression model fitted to the first half of the data is used to predict the second half. Typically the model does less well predicting the second half than it did predicting the first half, upon which it was based. The first half predictions are overly optimistic, often strikingly so, because the model has been selected to fit the first half data.

There is no need to divide the data set into equal halves. These days it is more common to leave out one data point at a time, fit the model to the remaining points, and see how well the fitted model predicts the excluded point. The average of the prediction errors, each point being left out once, is the cross-validated measure of prediction error. See Stone (1974) or Geisser (1975) for a full description, and Wahba and Wold (1975) for a nice application.

This form of cross validation looks like the jackknife in the sense that data points are omitted one at a time. However, there is no obvious statistic $\hat{\theta}$ being jackknifed, and any deeper connection between the two data ideas has been firmly denied in the literature.

This chapter discusses cross validation, the jackknife and the bootstrap in the regression context given above. It turns out that all three ideas are closely connected in theory, though not necessarily in their practical consequences. The concept of "excess error", vaguely suggested above, is formally defined in § 7.1. (In a discriminant analysis, for example, excess error is the difference between the true and apparent rate of classification errors.) The bootstrap estimate of excess error is easily obtained. Then the jackknife approximation to the bootstrap estimate is derived, and seen to be closely related to the cross-validated estimate.

7.1. Excess error. In a regression problem the data consists of pairs $(T_1, Y_1), (T_2, Y_2), \cdots, (T_n, Y_n)$, where T_i is a $1 \times p$ vector of predictors and Y_i is a real-valued response variable. In this discussion we take the simpler point of view mentioned at the end of § 5.7, that the $X_i = (T_i, Y_i)$ can be thought of as independent random quantities from an unknown distribution F on $\mathscr{X} = \mathscr{R}^{p+1}$,

$$(7.1.) \qquad X_1, X_2, \cdots, X_n \overset{\text{iid}}{\sim} F.$$

We observe $X_1 = x_1, X_2 = x_2, \cdots, X_n = x_n$, and denote $\mathbf{X} = (X_1, X_2, \cdots, X_n)$, $\mathbf{x} = (x_1, x_2, \cdots, x_n)$.

Having observed $\mathbf{X} = \mathbf{x}$, we have in mind some method of fitting a regression surface, which will then be used to predict future values of the response variable, say

$$(7.2.) \qquad \text{predicted value of } y_0 = \eta_{\mathbf{x}}(t_0).$$

(The subscript 0 indicates a new point $x_0 = (t_0, y_0)$ distinct from $x_1, x_2, \cdots, x_n$.) For example, ordinary linear regression fits the surface $\eta_{\mathbf{x}}(t_0) = t_0\hat{\beta}$, where $\hat{\beta} = (\mathbf{t}'\mathbf{t})^{-1}\mathbf{t}'\mathbf{y}$, $\mathbf{t}' = (t_1', t_2', \cdots, t_n')$, $\mathbf{y} = (y_1, y_2, \cdots, y_n)'$. Logistic regression, in which the y_i all equal 0 or 1, fits the surface $\eta_{\mathbf{x}}(t_0) = [1 + e^{-t_0\hat{\beta}}]^{-1}$, where $\hat{\beta}$ maximizes the likelihood function $\prod_{i=1}^{n} [e^{-t_i\beta(1-y_i)}/(1+e^{-t_i\beta})]$.

Section 7.6 gives an example of a much more elaborate fitting procedure, involving sequential decisions and a complicated model building process. Cross validation and the bootstrap are unfazed by such complications. The only restriction we impose is that $\eta_{\mathbf{x}}(\cdot)$ be a functional statistic, which means that it depends on $\mathbf{x}$ through $\hat{F}$, the empirical distribution (2.2); in other words, there exists $\eta(t_0, F)$, not depending on n, such that

$$(7.3) \qquad \eta_{\mathbf{x}}(t_0) = \eta(t_0, \hat{F}).$$

All of the common fitting methods, including linear regression and logistic regression, satisfy (7.3). In fact we only need (7.3) to establish the connection between the bootstrap and the other methods. The bootstrap itself requires the weaker condition that $\eta_{\mathbf{x}}(\cdot)$ be symmetrically defined in $x_1, x_2, \cdots, x_n$, and similarly for cross-validation.

Let $Q[y, \eta]$ be a measure of *error* between an observed value y and a prediction η. In ordinary regression theory the common measure is $Q[y, \eta] = (y-\eta)^2$. For logistic regression, in which we are trying to assign probabilities η to dichotomous events y, a familiar choice is

$$Q[y, \eta] = \begin{cases} 0 & \text{if } y = 1, \eta > \frac{1}{2} \quad \text{or} \quad y = 0, \eta \leqq \frac{1}{2}, \\ 1 & \text{otherwise.} \end{cases}$$

If events $y_1, y_2, \cdots, y_n$ have been assigned probabilities $\eta_1, \eta_2, \cdots, \eta_n$, then $\sum Q[y_i, \eta_i]$ is the number of prediction errors. Efron (1978) discusses other Q functions for this situation.

We will be interested in estimating a quantity called the "expected excess error". Define excess error as the random variable

$$(7.4) \qquad R(\mathbf{X}, F) = E_{0F}Q[Y_0, \eta_{\mathbf{X}}(T_0)] - E_{0\hat{F}}Q[Y_0, \eta_{\mathbf{X}}(T_0)].$$

The symbol "E_{0F}" indicates expectation over a single new point

$$X_0 = (T_0, Y_0) \sim F, \tag{7.5}$$

independent of $X_1, X_2, \cdots, X_n \overset{\text{iid}}{\sim} F$, the data which determine $\eta_{\mathbf{X}}(\cdot)$. ($\mathbf{X}$ is called the "training set" in the discrimination literature.) Likewise "$E_{0\hat{F}}$" indicates expectation over

$$X_0 \sim \hat{F} \tag{7.6}$$

independent of $\mathbf{X}$. Neither E_{0F} nor $E_{0\hat{F}}$ averages over $\mathbf{X}$, which is why $R(\mathbf{X}, F)$ is written as a function of $\mathbf{X}$. It is a function of F through the term $E_{0F}Q$.

The second term on the right of (7.4) equals

$$E_{0\hat{F}}Q[Y_0, \eta_{\mathbf{X}}(T_0)] = \frac{1}{n}\sum_{j=1}^{n} Q[Y_j, \eta_{\mathbf{X}}(T_j)],$$

since $\hat{F}$ puts mass $1/n$ at each point (T_j, Y_j). This is a statistic for which we observe the realization

$$\frac{1}{n}\sum_{j=1}^{n} Q[y_j, \eta_{\mathbf{x}}(t_j)] = \frac{1}{n}\sum_{j=1}^{n} Q[y_j, \hat{\eta}_j], \tag{7.7}$$

using the notation

$$\hat{\eta}_j = \eta_{\mathbf{x}}(t_j). \tag{7.8}$$

Statistic (7.7) is the "apparent error". Typically, since $\eta_{\mathbf{x}}(\cdot)$ is fitted to the observed data $\mathbf{x}$, this will be smaller than the "true error" $E_{0F}Q[Y_0, \eta_{\mathbf{x}}(T_0)]$, which is the expected error if $\eta_{\mathbf{x}}(\cdot)$ is used to predict a new Y_0 from its T_0 value. We are interested in estimating the expected excess error $E_F R(\mathbf{X}, F)$, the expected amount by which the true error exceeds the apparent error. A subtle point arises here: $E_F R$ is not the expected excess error for the regression surface $\eta_{\mathbf{x}}(\cdot)$ actually fitted, but rather the expectation over all potential regression surfaces $\eta_{\mathbf{X}}(\cdot)$. $E_F R$ is like the bias $E_F\hat{\theta} - \theta$, which is an average property of $\hat{\theta}(\mathbf{X})$, not of $\hat{\theta}(\mathbf{x}) - \theta$ for the particular $\mathbf{x}$ observed. This point is discussed again in § 7.4.

Example 7.1. *Linear discriminant analysis.* Suppose that in the training set y_j equals 1 or 2 as t_j comes from population 1 or population 2. For example, the t_j may be diagnostic variables on patients who do ($y_j = 1$) or don't ($y_j = 2$) require surgery. Given a new t_0 we wish to predict the corresponding y_0. Fisher's (estimated) linear discriminant function is

$$\eta_{\mathbf{x}}(t_0) = \hat{\alpha} + t_0\hat{\beta}, \tag{7.9}$$

where $\hat{\alpha}$ and $\hat{\beta}$ are calculated as follows. Let $n_1 = \#\{y_j = 1\}$, $n_2 = \#\{y_j = 2\}$,

$$\bar{t}_1 = \frac{1}{n_1}\sum_{y_j=1} t_j, \quad \bar{t}_2 = \frac{1}{n_2}\sum_{y_j=2} t_j, \quad \text{and} \quad S = \frac{1}{n}\left[\sum_{j=1}^{n} t_j t_j' - n_1\bar{t}_1'\bar{t}_1 - n_2\bar{t}_2'\bar{t}_2\right].$$

Then

$$\hat{\alpha} = \frac{[\bar{t}_1 S^{-1}\bar{t}_1' - \bar{t}_2 S^{-1}\bar{t}_2']}{2} \quad \text{and} \quad \hat{\beta} = (\bar{t}_2 - \bar{t}_1)S^{-1}. \tag{7.10}$$

The linear discriminant function (7.9) divides $\mathcal{R}^p$ into two sets

$$\mathcal{A}_1(\mathbf{x}) = \{t_0 \colon \eta_{\mathbf{x}}(t_0) \leqq 0\}, \qquad \mathcal{A}_2(\mathbf{x}) = \{t_0 \colon \eta_{\mathbf{x}}(t_0) > 0\}. \tag{7.11}$$

If $t_0 \in \mathcal{A}_2(x)$ then the prediction is made that $y_0 = 2$, while if $t_0 \in \mathcal{A}_1(x)$ the prediction is $y_0 = 1$. Why use this procedure? If the two populations are a priori equally likely,

$$\text{Prob}_F\{Y = 1\} = \text{Prob}_F\{Y = 2\} = \tfrac{1}{2}, \tag{7.12}$$

and if given $Y = y$, T is multivariate normal with mean vector μ_y and covariance matrix $\mathbf{\Sigma}$,

$$T|y \sim \mathcal{N}_p(\mu_y, \mathbf{\Sigma}), \tag{7.13}$$

then the linear discriminant estimates the Bayes classification procedure; see Efron (1975).

The apparent error in a discriminant analysis problem is the proportion of misclassified points in the training set,

$$\frac{\#\{j \colon t_j \in \mathcal{A}_1(\mathbf{x}),\ y_j = 2 \text{ OR } t_j \in \mathcal{A}_2(\mathbf{x}),\ y_j = 1\}}{n}. \tag{7.14}$$

In other words, we are using the error measure

$$Q[y, \eta] = \begin{cases} 0 & \text{if } y = 2,\ \eta > 0 \quad \text{or} \quad y = 1,\ \eta \leqq 0, \\ 1 & \text{otherwise.} \end{cases} \tag{7.15}$$

Suppose $p = 2$, $n = 14$, and the multivariate normal model (7.12), (7.13) is correct: $\mu_1 = -(\frac{1}{2}, 0)$, $\mu_2 = (\frac{1}{2}, 0)$, $\mathbf{\Sigma} = I$. A Monte Carlo experiment of 1000 trials showed that the expected apparent error was .262 (about $\frac{1}{4}$ of the 14 points misclassified) but that the expected true error was .356. In other words, the expected excess error $E_F R = .356 - .262 = .094$. This experiment is discussed further in § 7.4; see Table 7.1.

7.2. Bootstrap estimate of expected excess error. For convenience we denote an estimate of expected excess error by $\widehat{\text{EEE}}$. The bootstrap estimate is easy to write down. Consider a single bootstrap sample $\mathbf{X}^* = (X_1^*, X_2^*, \cdots, X_n^*)$, selected as in (5.5). The resampling vector $\mathbf{P}^* = (P_1^*, P_2^*, \cdots, P_n^*)$, $P_i^* = \#\{X_j^* = x_i\}/n$ has multinomial distribution (6.6). The bootstrap realization of random variable (7.4) is

$$R^* = R(\mathbf{X}^*, \hat{F}) = E_{0\hat{F}}Q[Y_0, \eta_{\mathbf{X}^*}(T_0)] - E_{0\hat{F}^*}Q[Y_0, \eta_{\mathbf{X}^*}(T_0)],$$

where $\hat{F}^*$ is the distribution putting mass P_i^* on x_i, and $\eta_{\mathbf{X}^*}(\cdot)$ is the regression surface determined by $\mathbf{X}^*$. Writing

$$\hat{\eta}_i^* = \eta_{\mathbf{X}^*}(t_i), \tag{7.16}$$

it is easy to see that

$$R^* = \sum_{i=1}^{n} (P_i^\circ - P_i^*)Q[y_i, \hat{\eta}_i^*], \tag{7.17}$$

$P_i^\circ = 1/n$ as before. The bootstrap estimate of expected excess error is

$$\widehat{\text{EEE}}_{\text{BOOT}} = E_* R^* = E_* \sum_{i=1}^{n} (P_i^\circ - P_i^*)Q[y_i, \hat{\eta}_i^*], \tag{7.18}$$

E_*R^* indicating expectation under bootstrap resampling (6.6). (Since the data set $\mathbf{x}$ is fixed, the $\hat{\eta}_i^*$ are functions of $\mathbf{P}^*$, under the assumption that $\eta_{\mathbf{X}^*}(\cdot)$ is symmetrically defined in $X_1^*, X_2^*, \cdots, X_n^*$.)

Example 7.2. *Ordinary linear regression.* As before, following (7.2), $\eta_{\mathbf{x}}(t_0) = t_0(\mathbf{t}'\mathbf{t})^{-1}\mathbf{t}'\mathbf{y}$. Define $\mathbf{D}_{\mathbf{P}^*}$ to be the diagonal matrix with $i$ith element P_i^*. Then

$$\eta_i^* = t_i(\mathbf{t}'\mathbf{D}_{\mathbf{P}^*}\mathbf{t})^{-1}(\mathbf{t}'\mathbf{D}_{\mathbf{P}^*}\mathbf{y}), \tag{7.19}$$

and $E_*R^* = E_*\sum_i (P_i^\circ - P_i^*)(y_i - \eta_i^*)^2$. Notice that $\widehat{\text{EEE}}_{\text{BOOT}} = E_*R^*$ is minus the covariance under (6.6) between P_i^* and $(y_i - \eta_i^*)^2$. Since large values of P_i^* tend to decrease $(y_i - \eta_i^*)^2$, it is clear that $\widehat{\text{EEE}}_{\text{BOOT}}$ should be positive.

7.3. Jackknife approximation to the bootstrap estimate. The excess risk random variable is of the form $R(\hat{F}, F)$, (6.33), under assumption (7.3). We can use (6.36) to get a jackknife approximation of E_*R^*. Since $R(\mathbf{P}^\circ) = 0$ in this case, (6.36) gives

$$E_*R^* \doteq \widehat{\text{EEE}}_{\text{JACK}} \equiv (n-1)R_{(\cdot)} = \frac{n-1}{n}\sum_{i=1}^{n} R(\mathbf{P}_{(i)}). \tag{7.20}$$

Define

$$\hat{\eta}_{(i)j} = \eta_{\mathbf{x}_{(i)}}(t_j), \tag{7.21}$$

where $\mathbf{x}_{(i)}$ is the data set $x_1, x_2, \cdots, x_{n-1}, x_{n-1}, \cdots, x_n$. Substituting $\mathbf{P}^* = \mathbf{P}_{(i)}$ in (7.17) gives

$$R(P_{(i)}) = \sum_{j=1}^{n} \left(\frac{1}{n} - P_{(i)j}\right) Q[y_j, \hat{\eta}_{(i)j}] = \frac{Q[y_i, \hat{\eta}_{(i)i}]}{n} - \frac{\sum_{j\neq i} Q[y_j, \hat{\eta}_{(i)j}]}{n(n-1)}$$

so

$$\begin{aligned}\widehat{\text{EEE}}_{\text{JACK}} &= \frac{n-1}{n^2}\sum_i Q[y_i, \hat{\eta}_{(i)i}] - \frac{1}{n^2}\sum_i \sum_{j\neq i} Q[y_j, \hat{\eta}_{(i)j}] \\ &= \frac{1}{n}\sum_i Q[y_i, \hat{\eta}_{(i)i}] - \frac{1}{n^2}\sum_i\sum_j Q[y_j, \hat{\eta}_{(i)j}] \\ &= \frac{1}{n}\sum_i Q[y_i, \hat{\eta}_{(i)i}] - \frac{1}{n}\sum_i \frac{\sum_j Q[y_i, \hat{\eta}_{(j)i}]}{n}.\end{aligned} \tag{7.22}$$

7.4. Cross-validation estimate of excess error. The cross-validation estimate of expected excess error is

$$\widehat{\text{EEE}}_{\text{CROSS}} = \frac{1}{n}\sum_i Q[y_i, \hat{\eta}_{(i)i}] - \frac{1}{n}\sum_i Q[y_i, \hat{\eta}_i], \tag{7.23}$$

the difference in observed error when we don't or do let x_i assist in its own prediction. Notice the similarity between (7.22) and (7.23). Before presenting any theoretical calculations, we give further Monte Carlo results from the experiment described at the end of § 7.1.

Table 7.1 shows the first 10 trials of the linear discriminant problem, true distributions normal as in (7.12), (7.13) and summary statistics for 100 and 1000 trials. The sample size is $n = 14$, dimension $p = 2$. We see that $\widehat{\text{EEE}}_{\text{JACK}}$ and $\widehat{\text{EEE}}_{\text{CROSS}}$ are almost the same, except in trial 2, with correlation .93 over 1000 trials. Neither method yields useful estimates. The values of $\widehat{\text{EEE}}$ are capriciously large or small, with coefficients of variations .073/.091 = .80 and .068/.093 = .73 in 100 trials. The bootstrap estimates, $B = 200$ bootstrap replications per trial, are much less variable, with coefficient of variation only .028/.080 = .35.

The actual excess error, $R(\mathbf{x}, F)$, is given in column A. It is quite variable from trial to trial. In 5 of the first 10 trials (and 22 of the first 100 trials) it is negative, the apparent error being *greater* than the true error. This is not a good situation for bias correction, i.e., for adjusting the apparent error rate by adding an estimate of expected excess error. The last column gives the bootstrap estimates of Sd(R) = .114,

$$\text{Sd}_* R^* = \left[\sum_{b=1}^{B} \frac{(R^{*b} - R^{*\cdot})^2}{B-1}\right]^{1/2}.$$

These estimates are seen to be quite dependable.

In Table 7.2 the situation is more favorable to bias correction. The actual excess error $R(\mathbf{x}, F)$ was positive in 98 out of 100 trials, averaging .184. $\widehat{\text{EEE}}_{\text{JACK}}$ and $\widehat{\text{EEE}}_{\text{CROSS}}$ are even more highly correlated, .98, and again both are too variable from trial to trial to be useful. The bootstrap estimates $\widehat{\text{EEE}}_{\text{BOOT}}$ are much less variable from trial to trial, but are biased downward. Adding $\widehat{\text{EEE}}_{\text{BOOT}}$ to the apparent error rate (7.14) substantially improves estimation of the true error rate in this case. The root mean square error of estimation for the true error rate decreases from .209 (using just (7.14)) to .145 (using (7.14) plus $\widehat{\text{EEE}}_{\text{BOOT}}$) for the first 10 trials. The comparable values for cross-validation and the jackknife are .148 and .151 respectively.

It would be nice if the estimates EEE correlated well with $R(\mathbf{x}, F)$, i.e., if the suggested bias corrections were big or small as the situation called for. In fact somewhat the opposite happens: the correlations are all negative, the bootstrap being most markedly so. Situations that produce grossly overoptimistic apparent errors, such as trial 10 of Table 7.2, tend to have the smallest estimated $\widehat{\text{EEE}}$. The author can't explain this phenomenon. Current research focuses on this problem, and on the downward bias of $\widehat{\text{EEE}}_{\text{BOOT}}$ evident in Table 7.2.

TABLE 7.1.

Expected excess error estimated by the bootstrap, by cross-validation and by the jackknife for the linear discriminant problem, $n = 14$, *true situation as in* (7.12), (7.13), $p = 2$, $\mu_1 = (-\frac{1}{2}, 0)$, $\mu_2 = (\frac{1}{2}, 0)$, $\Sigma = I$. *Results for first ten trials, summary statistics for first* 100 *trials and* 1000 *trials*. Corr. $(C, D) = .93$, Corr. $(A, C) = -.07$, Corr. $(A, D) = -.23$ (1000 *trials*); Corr. $(A, B) = -.64$ (100 *trials*).

Trial	n_1	Apparent error rate (7.14)	A Actual excess $R(\mathbf{x}, F)$	B Bootstrap estimate ($B = 200$) $\widehat{\text{EEE}}_{\text{BOOT}}$	C Cross-val estimate $\widehat{\text{EEE}}_{\text{CROSS}}$	D Jackknife estimate $\widehat{\text{EEE}}_{\text{JACK}}$	Bootstrap std dev estimate $\text{Sd}_* R^*$
1	9	.286	.172	.083	.214	.214	.117
2	6	.357	−.045	.098	.000	.066	.118
3	7	.357	−.044	.110	.071	.066	.108
4	8	.429	−.078	.107	.071	.066	.111
5	8	.357	−.027	.102	.143	.148	.120
6	8	.143	.175	.073	.214	.194	.094
7	8	.071	.239	.047	.071	.066	.077
8	6	.286	.094	.097	.071	.056	.109
9	7	.429	−.069	.127	.071	.087	.101
10	8	.143	.192	.048	.000	.010	.090
100 trials	Ave	.264	.096	.080	.091	.093	.104
	(Sd)	(.123)	(.114)	(.028)	(.073)	(.068)	(.014)
1000 trials	Ave	.262	.094		.097	.095	
	(Sd)	(.118)			(.085)	(.074)	

TABLE 7.2.
Same as Table 7.1, *except dimension* $p = 5$, $\mu_1 = (-1, 0, 0, 0, 0)$, $\mu_2 = (1, 0, 0, 0, 0)$. Corr. $(C, D) = .98$, Corr. $(A, C) = -.15$, Corr. $(A, D) = -.26$, Corr. $(A, B) = -.58$ (100 *trials*).

Trial	n_1	Apparent error rate (7.14)	A Actual excess $R(\mathbf{x}, F)$	B Bootstrap estimate ($B = 200$) $\widehat{\text{EEE}}_{\text{BOOT}}$	C Cross-Val estimate $\widehat{\text{EEE}}_{\text{CROSS}}$	D Jackknife estimate $\widehat{\text{EEE}}_{\text{JACK}}$	Bootstrap Std Dev estimate $\text{Sd}_* R^*$
1	7	.071	.135	.124	.357	.321	.112
2	5	.214	.010	.159	.357	.342	.102
3	8	.000	.247	.040	.000	.000	.064
4	6	.143	.098	.126	.143	.168	.098
5	6	.143	.101	.132	.286	.276	.090
6	4	.071	.229	.107	.143	.143	.106
7	8	.000	.236	.073	.143	.133	.070
8	8	.071	.142	.120	.357	.342	.082
9	5	.000	.269	.086	.214	.189	.068
10	8	.000	.239	.054	.071	.066	.080
100 trials	Ave	.069	.184	.103	.170	.167	.087
	(Sd)	(.076)	(.100)	(.031)	(.094)	(.089)	(.012)

7.5. Relationship between the cross-validation and jackknife estimates. Under reasonable conditions the expected excess error and the estimates $\widehat{\text{EEE}}_{\text{BOOT}}$, $\widehat{\text{EEE}}_{\text{JACK}}$ and $\widehat{\text{EEE}}_{\text{CROSS}}$ are of order of magnitude $O_p(1/n)$, while $\widehat{\text{EEE}}_{\text{JACK}} - \widehat{\text{EEE}}_{\text{CROSS}} = O_p(1/n^2)$. We briefly discuss the case of ordinary linear regression with quadratic error, $\eta_{\mathbf{x}}(t_0) = t_0(\mathbf{t}'\mathbf{t})^{-1}\mathbf{t}'\mathbf{y}$, $Q[y, \eta] = (y-\eta)^2$.

Define

$$r_i = y_i - \hat{\eta}_i \quad \text{and} \quad a_i = t_i(\mathbf{t}'\mathbf{t})^{-1}t_i'. \tag{7.24}$$

Under the usual regression assumptions r_i is $O_p(1)$, while a_i is $O_p(1/n)$. (Notice that $\sum_i a_i = \text{tr}\, I = p$, so by symmetry $E_F a_i = p/n$.) Using the matrix identity $[I - v'v]^{-1} = I + v'v/[1 - vv']$, v a row vector, we can express $\hat{\eta}_{(j)i} = t_i(\mathbf{t}'\mathbf{t} - t_j't_j)^{-1}(\mathbf{t}'\mathbf{y} - t_j'y_j)$ in a simple form,

$$\hat{\eta}_i - \hat{\eta}_{(j)i} = t_i(\mathbf{t}'\mathbf{t})^{-1}t_j'\frac{r_j}{1-a_j} = O_p\left(\frac{1}{n}\right). \tag{7.25}$$

In particular,

$$\hat{\eta}_i - \hat{\eta}_{(i)i} = \frac{a_i}{1-a_i} r_i. \tag{7.26}$$

Letting $\hat{\eta}_{(\cdot)i}$ denote $\sum_j \hat{\eta}_{(j)i}/n$, (7.25) gives

$$\begin{aligned} \hat{\eta}_i - \hat{\eta}_{(\cdot)i} &= t_i(\mathbf{t}'\mathbf{t})^{-1}\frac{1}{n}\sum_j t_j'\left(1 + \frac{a_j}{1-a_j}\right) r_j \\ &= t_i(\mathbf{t}'\mathbf{t})^{-1}\frac{1}{n}\sum_j t_j'\frac{a_j r_j}{1-a_j} = O_p\left(\frac{1}{n^2}\right). \end{aligned} \tag{7.27}$$

(We have used the orthogonality condition $\sum t_j' r_j = 0$.)

From (7.23),

$$\begin{aligned} \widehat{\text{EEE}}_{\text{CROSS}} &= \frac{1}{n}\sum_i [(r_i + \hat{\eta}_i - \hat{\eta}_{(i)i})^2 - r_i^2] \\ &= \frac{2}{n}\sum_i r_i(\hat{\eta}_i - \hat{\eta}_{(i)i}) + \frac{1}{n}\sum_i (\hat{\eta}_i - \hat{\eta}_{(i)i})^2 \\ &= \frac{2}{n}\sum_i \frac{a_i}{1-a_i} r_i^2 + \frac{1}{n}\sum_i \left(\frac{a_i}{1-a_i} r_i\right)^2 = O_p\left(\frac{1}{n}\right). \end{aligned}$$

Comparing (7.23) with (7.22) gives

$$\begin{aligned} \widehat{\text{EEE}}_{\text{JACK}} - \widehat{\text{EEE}}_{\text{CROSS}} &= \frac{1}{n}\sum_i (y_i - \hat{\eta}_i)^2 - \frac{1}{n}\sum_i (y_i - \hat{\eta}_{(\cdot)i})^2 \\ &= -\frac{2}{n}\sum_i r_i(\hat{\eta}_i - \hat{\eta}_{(\cdot)i}) + \frac{1}{n^2}\sum_i\sum_j (\hat{\eta}_i - \hat{\eta}_{(j)i})^2 \\ &= O_p\left(\frac{1}{n^2}\right). \end{aligned}$$

There is nothing special about ordinary linear regression in these calculations, except the tractability of the results. It seems likely that $(\widehat{\mathrm{EEE}}_{\mathrm{JACK}} - \widehat{\mathrm{EEE}}_{\mathrm{CROSS}})/\widehat{\mathrm{EEE}}_{\mathrm{CROSS}}$ is $O_p(1/n)$ under quite general conditions, but this remains to be proved. The results in Table 7.1, 7.2 are encouraging.

7.6. A complicated example. Figure 7.1 shows[1] a decision tree for classifying heart attack patients into low risk of dying (population 1) or high risk of dying (population 2) categories. A series of binary decisions brings a patient down the tree to a terminal node, which predicts either class 1 (e.g., node $T1$) or class 2 (e.g., node $T9$). For example, a patient with small PKCK value, small MNSBP value and finally large PKCK value ends up at $T3$, and is predicted to be in population 1. The numerical definition of "small" and "large" changes from node to node, so there is no contradiction between the first and third criterion of the previous sentence.

The decision tree, which is a highly nonlinear discriminant function, was based on a training set of 389 patients, 359 of whom survived their heart attacks at least 30 days (population 1), and 30 of whom didn't (population 2). Without going into details the tree's construction followed these rules:

1) The original 389 patients were divided into low and high groups using the best decision variable and the best dividing point for that variable. "Best" here means maximizing the separation between the two populations in a certain quantitative sense. (The ultimate best division would have all the "low" group in population 1 and all the "high" group in population 2, or vice-versa, in which case we could predict the training set perfectly on the basis of this one division, but of course that wasn't possible.) A computer search of all possible division points on each of 19 division variables selected PKCK, peak creatinine kinease level, to make the first division. Certain linear combinations of the variables, labeled F in Fig. 7.1, were also examined in the search for the best division.

2) Step 1 was repeated, separately, to best subdivide the high and low PKCK groups. (The low PKCK group was divided into low and high MNSBP groups; the high PKCK group into low and high RR groups.) The process was iterated, yielding a sequence of "complete" trees, the kth of which had 2^k terminal nodes. All subtrees of the complete trees were also considered. For example, Fig. 7.1 is a subtree of the 10th complete tree; it has 21 terminal nodes rather than the complete set of 2^{10}.

3) A terminal node was said to predict population 1 if $n_1/n_2 \geqq 8$, where n_i was the number of members of population i, in the training sample, at that node. If $n_1/n_2 < 8$ the prediction was population 2.

4) The tree in Fig. 7.1 predicts 41 population 1 patients into population 2 (nodes $T2$, $T6$, $T7$, $T9$, $T11$, $T13$, $T15$, $T17$, $T19$) and 1 population 2 patient into population 1 (node $T4$). The apparent error rates are $41/359 = 11.5\%$ for

[1] These data come from the Special Center of Research on Ischemic Heart Disease, University of California at San Diego, investigators John Ross, Jr., Elizabeth Gilpin, Richard Olshen and H. E. Henning, University of British Columbia. The tree was constructed by R. Olshen, who also performed the bootstrap analysis. It is a small part of a more extensive investigation.

population 1, 1/30 = 3.33% for population 2 and 42/389 = 10.8% overall. This tree was selected as best because it minimized the quantity {overall apparent error rate $+k \cdot$ number of terminal nodes}, k a certain constant.

These rules are not ad hoc; they are based on considerable theoretical work (see Gordon and Olshen (1978)). On the other hand, they are far too complicated for standard analysis. Instead, a bootstrap analysis was run, as in § 7.2. Only $B = 3$ bootstrap replications of (7.17) were generated, but these agreed closely with each other: the estimate $\widehat{\text{EEE}}_{\text{BOOT}}$ equaled 6.1%, making the bias corrected estimate of error rate 10.8% + 6.1% = 16.9%. More seriously, the bias corrected estimated error rate for the population 2 patients was 30%, compared to the apparent error rate of 3.33%! The tree in Fig. 7.1 does not predict population 2 patients, those with a high risk of dying, nearly as well as it appears to.

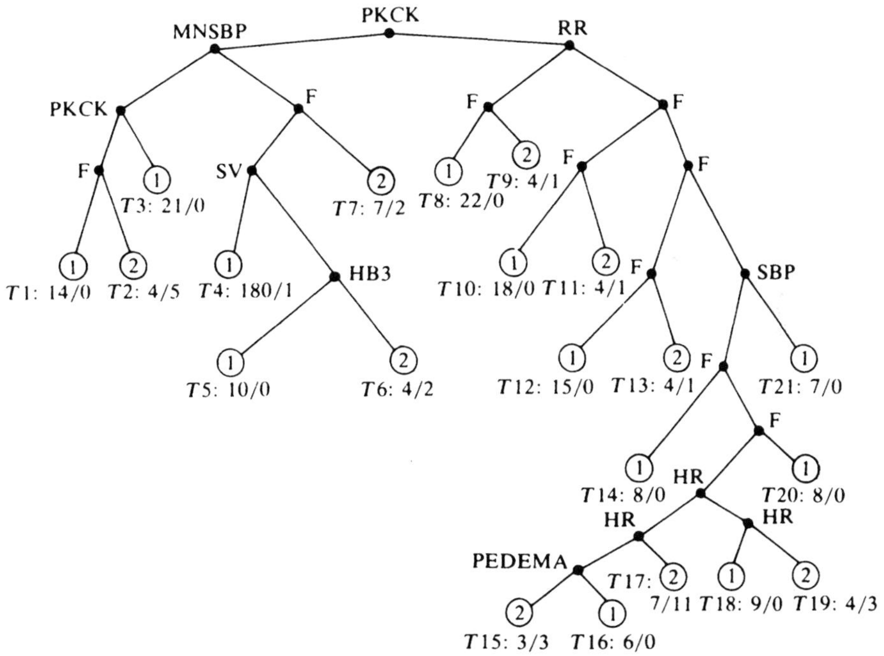

FIG. 7.1. *A decision tree for classifying heart attack patients into low risk of dying (population* 1) *or high risk of dying (population* 2). *Smaller values of the decision variables go to the left. Circled numbers at terminal nodes indicate population prediction. For example,* 6 *of the* 389 *patients in the training set end up at* T6, 4 *from population* 1, 2 *from population* 2; *these patients would all be predicted to be in population* 2.

Abbreviations. PKCK—*peak creatinine kinease level*; MNSBP—*minimum systolic blood pressure*; SBP—*systolic blood pressure*; RR—*respiration rate*; HR—*average heart rate*; SV—*superventricular arrhythmia*; HB3—*heart block 3rd degree*; PEDEMA—*peripheral edema*; F—*Fisher linear discriminant function, differing from node to node.*

CHAPTER 8

Balanced Repeated Replications (Half-sampling)

Half-sampling methods come from the literature of sampling theory. The basic idea is almost identical to the bootstrap estimate of standard deviation, but with a clever shortcut method, balanced repeated replications, that we haven't seen before. Kish and Frankel (1974) give a thorough review of the relevant sample survey theory.

In sampling theory it is natural to consider stratified situations where the sample space $\mathscr{X}$ is a union of disjoint strata $\mathscr{X}_h$,

$$\mathscr{X} = \bigcup_{h=1}^{H} \mathscr{X}_h.$$

(For example, $\mathscr{X}$ = United States, and $\mathscr{X}_h$ = state h, $h = 1, 2, \cdots, 50$.) The data consist of separate i.i.d. samples from each stratum,

$$X_{hi} \overset{\text{iid}}{\sim} F_h, \qquad i = 1, 2, \cdots, n_h, \quad h = 1, 2, \cdots, H, \tag{8.1}$$

where F_h is an unknown probability distribution on $\mathscr{X}_h$. Having observed $X_{hi} = x_{hi}$, $i = 1, \cdots, n_h$, $h = 1, \cdots, H$, define

$$\hat{F}_h: \text{mass } \frac{1}{n_h} \quad \text{on } x_{hi}, \tag{8.2}$$

the empirical probability distribution for stratum h, $h = 1, 2, \cdots, H$.

The goal of half-sampling theory is to assign an estimate of standard deviation to a functional statistic

$$\hat{\theta} = \theta(\hat{F}_1, \hat{F}_2, \cdots, \hat{F}_H). \tag{8.3}$$

For example $\hat{\theta}$ might be a linear functional statistic

$$\hat{\theta} = \sum_{h=1}^{H} \left\{ \mu_h + \frac{1}{n_h} \sum_{i=1}^{n_h} \alpha_h(X_{hi}) \right\}, \tag{8.4}$$

μ_h and $\alpha_h(\cdot)$ known, though possibly different for different strata. As another example, suppose $\mathscr{X}_h = \mathscr{R}^2$, $h = 1, \cdots, H$, and that π_h are known probabilities, $\sum_{h=1}^{H} \pi_h = 1$. The *mixture* $\sum_h \pi_h \cdot \hat{F}_h$ is a distribution on $\mathscr{R}^2$ putting mass π_h/n_h on each x_{hi}. In particular, if $\pi_h = n_h/n$ then $\sum_h \pi_h \cdot \hat{F}_h = \hat{F}$, (2.2). The sample correlation can be written $\rho(\hat{F}) = \rho\,(\sum_h n_h/n \cdot \hat{F}_h)$. In this way any functional statistic $\hat{\theta} = \theta(\hat{F})$ can be written in form (8.3). In the case of the correlation, we might prefer the statistic $\rho(\sum_h (n_h/n)\cdot \tilde{F}_h)$, where $\tilde{F}_h$ puts mass $1/n_h$ on $x_{hi}-\bar{x}_h$, $\bar{x}_h=\sum_i x_{hi}/n_h$. This is of form (8.3), but not of form (2.2).

8.1. Bootstrap estimate of standard deviation. The obvious bootstrap estimate of Sd $(\hat{\theta})$ is obtained, as in (5.4)–(5.6), by the following algorithm:

1. Construct the $\hat{F}_h$, (8.2).
2. Draw independent bootstrap samples $X^*_{hi} \overset{\text{iid}}{\sim} \hat{F}_h$, $i=1,\cdots,n_h$, $h=1,\cdots,H$; let $\hat{F}^*_h$ be the distribution putting mass P^*_{hi} on x_{hi}, where $P^*_{hi}=\#\{X^*_{hj}=x_{hi}\}/n_h$; and let $\hat{\theta}^*=\theta(\hat{F}^*_1,\hat{F}^*_2,\cdots,\hat{F}^*_H)$.
3. Independently repeat step 2, B times, obtaining bootstrap replications $\hat{\theta}^{*1},\hat{\theta}^{*2},\cdots,\hat{\theta}^{*B}$, and estimate Sd $(\hat{\theta})$ by (5.6),

$$\widehat{\text{SD}}_{\text{BOOT}}=\left\{\frac{\sum_{b=1}^{B}[\hat{\theta}^{*b}-\hat{\theta}^{*\cdot}]^2}{B-1}\right\}^{1/2}. \tag{8.5}$$

As before, $\widehat{\text{SD}}_{\text{BOOT}}$ is really defined as the limit of (8.5) as $B\to\infty$, but in most cases we have to settle for some finite value like $B=50$ or 100.

In the case of a linear statistic (8.4), where we can take $B=\infty$ without actually using Monte Carlo sampling, standard theory shows that

$$\widehat{\text{SD}}_{\text{BOOT}}=\left[\sum_{h=1}^{H}\frac{\hat{\sigma}_h^2}{n_h}\right]^{1/2}, \tag{8.6}$$

$\hat{\sigma}_h^2$ being the hth sample variance

$$\hat{\sigma}_h^2=\sum_{i=1}^{n_h}\frac{[\alpha_{hi}-\alpha_{h\cdot}]^2}{n_h}, \tag{8.7}$$

$\alpha_{hi}=\alpha_h(x_{hi})$, $\alpha_{h\cdot}=\sum_i \alpha_{hi}/n_h$. This compares with the true standard deviation

$$\text{Sd}(\hat{\theta})=\left[\sum_{h=1}^{H}\frac{\sigma_h^2}{n_h}\right]^{1/2}, \qquad \sigma_h^2=\text{Var}\,\alpha_h(X_{hi}). \tag{8.8}$$

8.2. Half-sample estimate of standard deviation. The expected value of $\widehat{\text{VAR}}_{\text{BOOT}}=\widehat{\text{SD}}^2_{\text{BOOT}}$ equals

$$E_{F_1,F_2,\cdots,F_H}\sum_h\frac{\hat{\sigma}_h^2}{n_h}=\sum_h\frac{n_h-1}{n_h}\frac{\sigma_h^2}{n_h}, \tag{8.9}$$

compared to the true value Var $\hat{\theta}=\sum\sigma_h^2/n_h$. Previously we have ignored the downward bias in (8.9), but in sampling theory the n_h are often small, and the bias can be severe. In particular, if all the $n_h=2$, the case most often considered in half-sampling theory, then $E\,\widehat{\text{VAR}}_{\text{BOOT}}=\frac{1}{2}\,\text{Var}\,(\hat{\theta})$.

The *half-sample*, or *repeated replications*, estimate of standard deviation, $\widehat{\text{SD}}_{\text{HS}}$, is the same as the bootstrap estimate, except that at step 2 of the algorithm we choose samples of size n_h-1 instead of n_h: $X^*_{hi}\overset{\text{iid}}{\sim}\hat{F}_h$, $i=1,2,\cdots,n_h-1$, $h=1,\cdots,H$. Reducing the size of the bootstrap samples by 1 removes the bias in $\widehat{\text{VAR}}$ when $\hat{\theta}$ is linear, since then

$$\widehat{\text{SD}}_{\text{HS}} = \left[\sum_{h=1}^{H} \frac{\hat{\sigma}_h^2}{n_h - 1} \right]^{1/2},$$

and so $E\ \widehat{\text{VAR}}_{\text{HS}} = \sum_h \sigma_h^2/n_h = \text{Var}\ \hat{\theta}$.

Suppose all the $n_h = 2$. Then the half-sample method really chooses half-samples. One of the two data points from each stratum is selected to be the bootstrap observation, independently and with equal probability,

$$(8.10)\qquad X_{h1}^* = \begin{Bmatrix} x_{h1} \\ x_{h2} \end{Bmatrix} \text{Prob}_* \tfrac{1}{2}, \quad \text{independently}, \quad h = 1, 2, \cdots, H.$$

In other words, each $\hat{F}_h^*$ is a one-point distribution putting all of its mass at either x_{h1} or x_{h2}, with equal probability.

Henceforth we will only discuss the situation where $n_h = 2$ for $h = 1, 2, \cdots, H$. There are $n = 2H$ observations in this case and 2^H possible half-samples. Let $\boldsymbol{\varepsilon} = (\varepsilon_1, \varepsilon_2, \cdots, \varepsilon_H)$ be a vector of ± 1's, indicating a half-sample according to the rule

$$(8.11)\qquad X_{h1}^* = \begin{cases} x_{h1} & \text{if } \varepsilon_h = +1, \\ x_{h2} & \text{if } \varepsilon_h = -1. \end{cases}$$

The set $\mathcal{J}_0$ of all possible vectors $\boldsymbol{\varepsilon}$ has $J_0 = 2^H$ members, each of which is selected with equal probability under half-sampling. Since $\boldsymbol{\varepsilon}$ determines all the $\hat{F}_h^*$, by (8.11), we can write $\hat{\theta}(\boldsymbol{\varepsilon})$ in place of $\hat{\theta}(\hat{F}_1^*, \hat{F}_2^*, \cdots, \hat{F}_H^*)$. In this notation, the half-sample estimate of standard deviation is

$$(8.12)\qquad \widehat{\text{SD}}_{\text{HS}} = \left\{ \sum_{\boldsymbol{\varepsilon} \in \mathcal{J}_0} \frac{[\hat{\theta}(\boldsymbol{\varepsilon}) - \hat{\theta}(\cdot)]^2}{J_0} \right\}^{1/2},$$

where $\hat{\theta}(\cdot) = \sum_{\boldsymbol{\varepsilon} \in \mathcal{J}_0} \hat{\theta}(\boldsymbol{\varepsilon})/J_0$.

Notice that J_0 is not what we called B before. In fact $B = \infty$, since we have considered all 2^H possible outcomes of $\hat{F}_1^*, \hat{F}_2^*, \cdots, \hat{F}_H^*$. That is why we divide by J_0 rather than $J_0 - 1$ in (8.12).

Line 8 of Table 5.2 shows half-sampling applied to the correlation coefficient $\hat{\rho}$, and to $\hat{\phi} = \tanh^{-1} \hat{\rho}$. The strata were defined artificially, (x_1, x_2) representing stratum 1, (x_3, x_4) stratum 2, $\cdots$, (x_{13}, x_{14}) stratum 7. For each of the 200 Monte Carlo trials, all 128 half-sample values were evaluated, and $\widehat{\text{SD}}_{\text{HS}}$ calculated according to (8.12). The numerical results are discouraging. Both bias and root mean square error are high, for both $\widehat{\text{SD}}_{\text{HS}}(\hat{\rho})$ and $\widehat{\text{SD}}_{\text{HS}}(\hat{\phi})$. Of course this is not a naturally stratified situation, so there is no particular reason to do half-sampling. On the other hand, it would be nice if the method worked well since, as we shall see, it can be implemented with considerably less computation than the bootstrap.

In line 10 of Table 5.2, the Sd estimates for each of the 200 trials were constructed using 128 randomly selected (out of all $14!/(7!)^2$ possible) half-samples. This method removes the component of variance in $\widehat{\text{SD}}_{\text{HS}}$ due to the artificial creation of strata, but the numerical results are still poor compared to the bootstrap results of line 1.

8.3. Balanced repeated replications. Suppose that $\mathcal{J}$ is a subset of $\mathcal{J}_0$ containing J vectors $\boldsymbol{\varepsilon}$, say $\mathcal{J}=\{\boldsymbol{\varepsilon}^1, \boldsymbol{\varepsilon}^2, \cdots, \boldsymbol{\varepsilon}^J\}$, and that these vectors satisfy

$$\sum_{j=1}^{J} \varepsilon_h^j \varepsilon_k^j = 0, \qquad 1 < h < k \leqq H. \tag{8.13}$$

McCarthy (1969) calls $\mathcal{J}$ a *balanced set* of half-samples. We will also require that a balanced set satisfy

$$\sum_{j=1}^{J} \varepsilon_h^j = 0, \qquad 1 \leqq h \leqq H. \tag{8.14}$$

The complete set $\mathcal{J}_0$ is itself balanced.

We define the *balanced half-sample* estimate[1] of standard deviation

$$\widehat{\text{SD}}_{\text{BHS}} = \left\{\frac{1}{J}\sum_{j=1}^{J} [\hat\theta(\boldsymbol{\varepsilon}^j) - \hat\theta(\cdot)]^2\right\}^{1/2}, \tag{8.15}$$

$\hat\theta(\cdot) = \sum_{j=1}^{J} \hat\theta(\boldsymbol{\varepsilon}^j)/J$. This is McCarthy's method of *balanced repeated replications*, and has the advantage of requiring only J instead of $J_0 = 2^H$ recomputations of $\hat\theta$, while still giving the same Sd estimate for linear statistics.

THEOREM 8.1. (McCarthy). *For a linear statistic* (8.4), $\widehat{\text{SD}}_{\text{BHS}} = \widehat{\text{SD}}_{\text{HS}}$.

Proof. With $\alpha_{hi} = \alpha_h(x_{hi})$, $\alpha_{h\cdot} = (\alpha_{h1} + \alpha_{h2})/2$ as before,

$$\hat\theta(\boldsymbol{\varepsilon}^j) = \sum_{h=1}^{H}\left\{\mu_h + \alpha_{h\cdot} + \varepsilon_h^j \frac{(\alpha_{h1} - \alpha_{h2})}{2}\right\} = \hat\theta + \sum_h \varepsilon_h^j \frac{(\alpha_{h1} - \alpha_{h2})}{2}, \tag{8.16}$$

so $\hat\theta(\cdot) = \sum_j \hat\theta(\boldsymbol{\varepsilon}^j)/J = \hat\theta$ because of (8.14). Then

$$\begin{aligned}\widehat{\text{SD}}_{\text{BHS}} &= \left\{\frac{1}{J}\sum_{j=1}^{J}\left[\sum_{h=1}^{H} \varepsilon_h^j \frac{(\alpha_{h1} - \alpha_{h2})}{2}\right]^2\right\}^{1/2} \\ &= \left\{\frac{1}{J}\sum_{h=1}^{H}\sum_{k=1}^{H}\sum_{j=1}^{J} \varepsilon_h^j\varepsilon_k^j \frac{(\alpha_{h1} - \alpha_{h2})(\alpha_{k1} - \alpha_{k2})}{4}\right\}^{1/2} \\ &= \left\{\sum_{h=1}^{H}\left(\frac{\alpha_{h1} - \alpha_{h2}}{2}\right)^2\right\}^{1/2},\end{aligned} \tag{8.17}$$

using (8.13). This last expression doesn't depend on $\mathcal{J}$, so $\widehat{\text{SD}}_{\text{BHS}}$ for linear statistics is the same for all balanced sets, including $\mathcal{J}_0$, which proves the theorem. □

[1] McCarthy's result is stated with $\hat\theta$ replacing $\hat\theta(\cdot)$ in (8.12) and (8.15), in which case condition (8.14) is not required. This replacement makes almost no difference in Table 5.2, and probably not in most cases, but if there were a substantial difference, definitions (8.12), (8.15) would be preferred for estimating standard deviation. McCarthy's definition is more appropriate for estimating root mean square error.

Line 10 of Table 5.2 gives summary statistics for $\widehat{\text{SD}}_{\text{BHS}}$ applied to the correlation experiment. A balanced set $\mathcal{J}$ with $J=8$ members was used, so each $\widehat{\text{SD}}_{\text{BHS}}$ required only 8 recomputations of $\hat{\rho}$ (or $\hat{\phi}$), instead of 128 as in line 8. The vectors $\boldsymbol{\varepsilon}^j$ were the rows of the matrix

$$(8.18)\qquad \begin{pmatrix} 1 & -1 & -1 & 1 & -1 & 1 & 1 \\ 1 & 1 & -1 & -1 & 1 & -1 & 1 \\ 1 & 1 & 1 & -1 & -1 & 1 & -1 \\ -1 & 1 & 1 & 1 & -1 & -1 & 1 \\ 1 & -1 & 1 & 1 & 1 & -1 & -1 \\ -1 & 1 & -1 & 1 & 1 & 1 & -1 \\ -1 & -1 & 1 & -1 & 1 & 1 & 1 \\ -1 & -1 & -1 & -1 & -1 & -1 & -1 \end{pmatrix}.$$

The results are quite similar to those of line 8, as Theorem 8.1 would suggest, although $\sqrt{\text{MSE}}$ is somewhat increased.

8.4. Complementary balanced half-samples. The *complementary half-sample* to that represented by $\boldsymbol{\varepsilon}$ is $-\boldsymbol{\varepsilon}$, i.e., the other half of the data. Suppose now that a balanced set $\mathcal{J}$ is also *closed under complementation*, so that if $\boldsymbol{\varepsilon}\in\mathcal{J}$ then $-\boldsymbol{\varepsilon}\in\mathcal{J}$. Then J is even, and we can index $\mathcal{J}$ so that each $\boldsymbol{\varepsilon}^j$, $j=1,2,\cdots,J/2$, is complementary to $\boldsymbol{\varepsilon}^{j+J/2}$. (In other words, the second half of $\mathcal{J}$ is complementary to the first half.) The complete balanced set $\mathcal{J}_0$ is closed under complementation.

The *complementary balanced half-sample* estimate of standard deviation is

$$(8.19)\qquad \widehat{\text{SD}}_{\text{CBHS}}=\left\{\frac{2}{J}\sum_{j=1}^{J/2}\left[\frac{\hat{\theta}(\boldsymbol{\varepsilon}^j)-\hat{\theta}(-\boldsymbol{\varepsilon}^j)}{2}\right]^2\right\}^{1/2}.$$

The advantage of $\widehat{\text{SD}}_{\text{CBHS}}$ is that Theorem 8.1 can now be extended to quadratic statistics. First we have to define what we mean by a quadratic statistic in the stratified context (8.1). The easiest definition is by analogy with (6.8). Let $\mathbf{P}^\circ=(\frac{1}{2},\frac{1}{2})$ and

$$(8.20)\qquad \mathbf{P}_h^j=\begin{cases}(1,0) & \text{if } \varepsilon_h^j=1,\\ (0,1) & \text{if } \varepsilon_h^j=-1.\end{cases}$$

A statistic $\hat{\theta}$ is *quadratic* if its half-sample values can be expressed as

$$(8.21)\qquad \hat{\theta}(\boldsymbol{\varepsilon}^j)=\hat{\theta}+\sum_{h=1}^{H}(\mathbf{P}_h^j-\mathbf{P}^\circ)\mathbf{U}_h+\frac{1}{2}\sum_{h=1}^{H}\sum_{k=1}^{H}(\mathbf{P}_h^j-\mathbf{P}^\circ)\mathbf{V}_{hk}(\mathbf{P}_k^j-\mathbf{P}^\circ)',$$

where the $\mathbf{U}_h$ are 1×2 vectors and the $\mathbf{V}_{hk}$ are 2×2 matrices. Quadratic functional statistics (4.14), (4.15) can be rewritten in this form. A linear functional statistic (8.4) is of form (8.21) with $\mathbf{V}_{hk}=\mathbf{0}$, $\mathbf{U}_h=(\alpha_{h1},\alpha_{h2})$.

THEOREM 8.2. *For a quadratic statistic* (8.21), $\widehat{\text{SD}}_{\text{CBHS}}$ *has the same value for all balanced sets* $\mathcal{J}$ *closed under complementation, including the complete set* $\mathcal{J}_0$.

Proof. $\mathbf{P}_h^j - \mathbf{P}^\circ = \varepsilon_h^j \mathbf{d}$, where $\mathbf{d} = (\frac{1}{2}, -\frac{1}{2})$, so

$$\frac{\hat{\theta}(\boldsymbol{\varepsilon}^j) - \hat{\theta}(-\boldsymbol{\varepsilon}^j)}{2} = \sum_{h=1}^{H} \varepsilon_h^j \mathbf{d}\mathbf{U}_h = \sum_{h=1}^{H} \varepsilon_h^j \frac{U_{h1} - U_{h2}}{2}, \tag{8.22}$$

the quadratic terms in (8.21) canceling out. The same calculation as in the proof of Theorem 8.1 shows that

$$\widehat{\mathrm{SD}}_{\mathrm{CBHS}} = \left\{ \sum_{h=1}^{H} \left[\frac{U_{h1} - U_{h2}}{2} \right]^2 \right\}^{1/2}. \tag{8.23}$$

□

Line 11 of Table 5.2 refers to $\widehat{\mathrm{SD}}_{\mathrm{CBHS}}$ for $\mathcal{J} = \mathcal{J}_0$, the complete set of 128 half-samples. Line 12 refers to $\widehat{\mathrm{SD}}_{\mathrm{CBHS}}$ for $\mathcal{J}$ consisting of 16 half-samples, the 8 displayed in (8.18) plus their complements. The results shown in lines 11 and 12 are remarkably similar, much more so than lines 8 and 10. Root mean square error is reduced, compared to $\widehat{\mathrm{SD}}_{\mathrm{HS}}$, although the results are still disappointing compared to the bootstrap, especially for $\hat{\phi}$.

The averages for $\widehat{\mathrm{SD}}_{\mathrm{CBHS}}$ shown in Table 5.2 are smaller than those for $\widehat{\mathrm{SD}}_{\mathrm{BHS}}$. This must always be the case.

THEOREM 8.3. *For any statistic $\hat{\theta}$ and any balanced set $\mathcal{J}$ closed under complementation,* $\widehat{\mathrm{SD}}_{\mathrm{BHS}} \geqq \widehat{\mathrm{SD}}_{\mathrm{CBHS}}$.

Proof. From definition (8.15),

$$\widehat{\mathrm{SD}}^2_{\mathrm{BHS}} = \frac{1}{J} \sum_{j=1}^{J} [\hat{\theta}(\boldsymbol{\varepsilon}^j) - \hat{\theta}(\cdot)]^2 = \frac{2}{J} \sum_{j=1}^{J/2} \frac{[\hat{\theta}(\boldsymbol{\varepsilon}^j) - \hat{\theta}(\cdot)]^2 + [\hat{\theta}(-\boldsymbol{\varepsilon}^j) - \hat{\theta}(\cdot)]^2}{2}$$

$$\geqq \frac{2}{J} \sum_{j=1}^{J/2} \left[\frac{\hat{\theta}(\boldsymbol{\varepsilon}^j) - \hat{\theta}(-\boldsymbol{\varepsilon}^j)}{2} \right]^2 = \widehat{\mathrm{SD}}^2_{\mathrm{CBHS}}.$$

Here we have used the elementary inequality $(a^2 + b^2)/2 \geqq [(a-b)/2]^2$. □

8.5. Some possible alternative methods. The half-sample form of the bootstrap, in which each stratum's bootstrap sample size is reduced by 1, is not the only way to correct the bias in the linear case. Still considering only the situation where all $n_h = 2$ we could, for example, simply multiply formula (8.5) by $\sqrt{2}$, i.e., estimate Sd by $\sqrt{2}\, \mathrm{SD}_{\mathrm{BOOT}}$.

Even if we wish to use half-sampling, we might prefer to half-sample from distributions other than $\hat{F}_h$, (8.2). Suppose, for instance, that $\mathcal{X}_h = \mathcal{R}^1$, $h = 1, 2, \cdots, H$, and let $\hat{\mu}_h = (x_{h1} + x_{h2})/2$, $\hat{\sigma}_h^2 = (x_{h1} - \hat{\mu}_h)^2$, the sample mean and variance of $\hat{F}_h$. Define

$$\tilde{F}: \text{mass } \frac{1}{n} \text{ at } \frac{x_{hi} - \hat{\mu}_h}{\hat{\sigma}_h}, \quad i = 1, 2, \quad h = 1, \cdots, H.$$

and let $\tilde{F}_h$ be the distribution of $\hat{\mu}_h + \hat{\sigma}_h \tilde{X}$, where $\tilde{X} \sim \tilde{F}$. Then $\tilde{F}_h$ has the same mean and variance as $\hat{F}_h$, but makes use of information from all strata to estimate the distribution in stratum h. Half-sampling from the $\tilde{F}_h$, (i.e., independently selecting $X_{h1}^* \sim \tilde{F}_h$, $h = 1, 2, \cdots, H$ and computing the standard deviation of

$\hat{\theta}(\hat{F}_1^*, \hat{F}_2^*, \cdots, \hat{F}_H^*)$ where $\hat{F}_h^*$ puts all its mass on X_{h1}^*) might be better than half-sampling from the $\hat{F}_h$.

Balanced half-sampling and complementary balanced half-sampling are not the only ways to cut down the amount of computation needed to estimate a standard deviation. Let $\hat{\theta}^h = \hat{\theta}(x_{11}, x_{12}, x_{21}, x_{22}, \cdots, x_{h1}, x_{h1}, \cdots, x_{H1}, x_{H2})$, the value of the statistic when x_{h2} is replaced by a duplicate of x_{h1}, but no other changes are made in the data set; and let $\hat{\theta}^{-h}$ be the value when instead x_{h1} is replaced by a duplicate of x_{h2}. Then it can be shown that

$$\widehat{\text{SD}} = \left\{ \sum_{h=1}^{H} \left[\frac{\hat{\theta}^h - \hat{\theta}^{-h}}{2} \right]^2 \right\}^{1/2}$$

equals $\widehat{\text{SD}}_{\text{CBHS}}$, (8.23), for quadratic statistics, (8.21). Evaluating this $\widehat{\text{SD}}$ requires only n recomputations of $\hat{\theta}$, which is the minimum possible number required for $\widehat{\text{SD}}_{\text{CBHS}}$. The relationship between this $\widehat{\text{SD}}$ and $\widehat{\text{SD}}_{\text{CBHS}}$ is analogous to the relationship between $\widehat{\text{SD}}_{\text{JACK}}$ and $\widehat{\text{SD}}_{\text{BOOT}}$.

Looking in the other direction, the clever idea underlying balanced repeated replications might be extended to reduce the number of calculations necessary for $\widehat{\text{SD}}_{\text{BOOT}}$. Artificial stratification into pairs is not a good general answer, as we saw in Table 5.2, but more ambitious stratification schemes seem promising.

CHAPTER 9

Random Subsampling

Hartigan (1969) introduced another resampling plan which we will call *random subsampling*. It is designed to give exact confidence intervals, rather than just standard deviations, but in a special class of problems: that of estimating the center of a symmetric distribution on the real line. We begin with a description of the problem and Hartigan's "typical value theorem", which very neatly gives the desired confidence intervals. Then we go on to show the connection between random subsampling and the bootstrap, in terms of large sample theory. Chapter 10 concerns the important problem of small sample nonparametric confidence intervals for nonsymmetric problems.

9.1. *m*-estimates. We consider the case of i.i.d. observations from a symmetric distribution on the real line,

$$X_1, X_2, \cdots, X_n \overset{\text{iid}}{\sim} F_\theta, \qquad \text{Prob}_\theta\{X \in A\} = \int_A f(x-\theta)\,dx, \tag{9.1}$$

where $f(\cdot)$ is a symmetric density function, $\int_{-\infty}^{\infty} f(x)\,dx = 1, f(x) \geqq 0, f(-x) = f(x)$. The unknown translation parameter θ is the center of symmetry of F_θ.

An *m-estimate* $\hat\theta(x_1, x_2, \cdots, x_n)$ for θ is any solution to the equation

$$\sum_{i=1}^{n} \psi(x_i - t) = 0. \tag{9.2}$$

Here the observed data $X_i = x_i$, $i = 1, 2, \cdots, n$ are fixed while t is the variable. The kernel $\psi(\cdot)$ is assumed to be antisymmetric and strictly increasing:

$$\text{i)}\quad \psi(-z) = -\psi(z), \qquad \text{ii)}\quad \psi(z)\uparrow z. \tag{9.3}$$

This last condition is not usually imposed, but is necessary for the development which follows. It guarantees that $\hat\theta$ is defined uniquely for any data set $x_1, x_2, \cdots, x_n$. Notice that $\hat\theta$ is a functional statistic, $\hat\theta = \theta(\hat F)$, since (9.2) can be written as $\int_{-\infty}^{\infty} \psi(x-t)\,d\hat F(x) = 0$.

Example 9.1. $\psi(z) = z$. Then $\hat\theta = \bar x$, the sample mean.

Example 9.2. $\psi(z) = \text{sgn}(z)\,[1 - e^{-c|z|}]$ for some constant $c > 0$. As $c \to \infty$, $\hat\theta_c(x_1, x_2, \cdots, x_n) \to$ sample median, the middle order statistic if n is odd, the average of the middle two order statistics if n is even.

Example 9.3. $\psi(z) = -f'(z)/f(z) = -(d/dz)\log f(z)$. Then (9.2) says that $\hat\theta$ is the solution to $\sum_i (\partial/\partial t)\log f(x_i - t) = 0$, i.e., the maximum likelihood estimate of θ. (This is the origin of the name "m-estimator".) If $f(z)$ is the normal density then $\psi(z) = z$ and $\hat\theta = \bar x$. If $f(z) = \frac{1}{2}e^{-|z|}$, the double exponential, then $\psi(z) =$

sgn (z) and $\hat{\theta}$ = sample median. If $f(z)=(1/\pi)(1+z^2)^{-1}$, the Cauchy distribution, then $\psi(z)=2z/(1+z^2)$. In the last two examples, condition (ii) of (9.3) isn't satisfied.

The influence function (4.13) of an m-estimate based on $\psi(\cdot)$ is IF $(x)= c\psi(x-\theta)$, c some positive constant. Robustness theory focuses on choices of $\psi(\cdot)$ which have bounded influence functions, sup $|\psi(z)|<\infty$, but still give reasonably high estimation efficiency for standard families like the normal, double exponential and Cauchy. Huber (1974) gives a thorough review of this theory.

9.2. The typical value theorem. There are 2^n-1 nonempty subsets of $\{1, 2, \cdots, n\}$. If S is such a subset, define $\hat{\theta}_S$ as the m-estimate based on $\{x_i; i\in S\}$,

(9.4) $$\hat{\theta}_S\colon \sum_{i\in S} \psi(x_i-t)=0.$$

The 2^n-1 values of $\hat{\theta}_S$ will be distinct with probability one, under assumptions (9.3). Their ordered values partition the line into 2^n intervals, say $\mathcal{I}_1, \mathcal{I}_2, \cdots, \mathcal{I}_{2^n}$. For instance $\mathcal{I}_1=(-\infty, x_{(1)})$, where $x_{(1)}$ is the smallest value of $x_1, x_2, \cdots, x_n$.

TYPICAL VALUE THEOREM[1]. *The true value θ has probability $1/2^n$ of being in any interval $\mathcal{I}_j$, $j=1, 2, \cdots, 2^n$.*

Proof. There are 2^n vectors $\boldsymbol{\delta}=(\delta_1, \delta_2, \cdots, \delta_n)$ having components $\delta_i=\pm 1$. For each $\boldsymbol{\delta}$ define

(9.5) $$Q(\boldsymbol{\delta}, t)=\sum_{i=1}^{n} \psi(\delta_i(x_i-t))=\sum_{i=1}^{n} \delta_i\psi(x_i-t).$$

In particular $Q(\mathbf{1}, t)=\sum_{i=1}^{n} \psi(x_i-t)$, the defining function for $\hat{\theta}$ in (9.2). The quantity

(9.6) $$Q(\boldsymbol{\delta}, t)-Q(\mathbf{1}, t)=-2\sum_{i\in S(\boldsymbol{\delta})} \psi(x_i-t),$$

where

(9.7) $$S(\boldsymbol{\delta})=\{i\colon \delta_i=-1\}$$

is strictly increasing in t, if $\boldsymbol{\delta}\neq\mathbf{1}$.

The following two statements are seen to be equivalent:

(9.8) $$\text{i)}\quad Q(\boldsymbol{\delta}, \theta)>Q(\mathbf{1}, \theta), \qquad \text{ii)}\quad \hat{\theta}_{S(\boldsymbol{\delta})}<\theta,$$

(since by (9.6) $Q(\boldsymbol{\delta}, \theta)-Q(\mathbf{1}, \theta)>0$ implies that $\sum_{i\in S(\boldsymbol{\delta})} \psi(x_i-t)$ has its root to the left of θ). But by the symmetry of $\psi(\cdot)$ and $f(\cdot)$ the 2^n random variables $Q(\boldsymbol{\delta}, \theta)=\sum_{i=1}^{n} \delta_i\psi(X_i-\theta)$ are exchangeable. Thus

(9.9) $$\text{Prob}_\theta\,\{Q(\mathbf{1}, \theta) \text{ is } j\text{th largest among the } Q(\boldsymbol{\delta}, \theta)\}=\frac{1}{2^n},$$

[1] From Hartigan (1969), who also credits J. Tukey and C. Mallows. The derivation here follows Maritz (1979).

$j = 1, 2, \cdots, 2^n$, so by (9.8),

$$\text{(9.10)} \qquad \text{Prob}_\theta\,\{\text{exactly } j-1 \text{ of the } \hat{\theta}_S < \theta\} = \frac{1}{2^n},$$

$j = 1, 2, \cdots, 2^n$, which is the statement of the theorem. □

The typical value theorem is used to set confidence intervals as follows: suppose we observe $n = 10$ observations from a symmetric density on the real line. Let $\hat{\theta}_{(1)} < \hat{\theta}_{(2)} < \cdots < \hat{\theta}_{(1023)}$ be the ordered sub-sample m-estimates. Then $(\hat{\theta}_{(51)}, \hat{\theta}_{(973)}) = \bigcup_{j=52}^{973} \mathcal{I}_j$ is a $922/1024 = .900$ central confidence interval for θ. *Warning*: an exact confidence interval is not necessarily a good one. For instance if $\psi(z) = z$, so $\hat{\theta} = \bar{x}$, and $f(z)$ is Cauchy, then $(\bar{x}_{(51)}, \bar{x}_{(973)})$ is a 90% central confidence interval for the center of the Cauchy distribution, $\bar{x}_{(j)}$ being the jth ordered subsample average. This interval can be absurdly long, compared to the optimum interval for the Cauchy if the sample includes an outlying observation.

9.3. Random subsampling. One needn't evaluate all $2^n - 1$ subsample values $\hat{\theta}_S$ in order to use the typical value theorem. *Random subsampling* provides a convenient shortcut.

COROLLARY. *Let $S_1, S_2, \cdots, S_{B-1}$ be chosen randomly and without replacement from the $2^n - 1$ nonempty subsets of $\{1, 2, \cdots, n\}$, and let $\mathcal{I}_1, \mathcal{I}_2, \cdots, \mathcal{I}_B$ be the intervals determined by the ordered values of $\hat{\theta}_{S_i}$. Then θ has probability $1/B$ of being in any interval $\mathcal{I}_j$, $j = 1, 2, \cdots, B$.*

The proof, which appears in Hartigan (1969), is left as a pleasant exercise for the reader.

Note. "Probability" has a different meaning in the corollary than in the theorem. Write $X_i = \theta + \varepsilon_i |X_i - \theta|$, so that

$$\text{(9.11)} \qquad \varepsilon_i \,\Big|\, |x_i - \theta| = \begin{Bmatrix} 1 \\ -1 \end{Bmatrix} \text{ prob } \tfrac{1}{2} \quad \text{independently}, \qquad i = 1, 2, \cdots, n,$$

"Probability" in the theorem refers to the conditional distribution (9.11) of the ε_i given the $|x_i - \theta|$. The probability statement in the corollary averages also over the random choice of the subsets S_j. This can be less satisfactory. Suppose, as a rather farfetched example, that $B = n$ and that we happen to choose $S_1 = \{x_1\}$, $S_2 = \{x_2\}, \cdots, S_n = \{x_n\}$. Then $\hat{\theta}_{S_i} = x_i$ for any m-estimate, and so $(x_{(1)}, x_{(n)})$ is an $(n-1)/(n+1)$ central confidence interval for θ, by the corollary. However, since θ is the median of the distribution generating the data $(x_{(1)}, x_{(n)})$ is a $1 - (\frac{1}{2})^{n-1}$ central confidence interval for θ! (See § 10.2.)

Hartigan also provides a more satisfactory version of the corollary, in which "probability" refers only to mechanism (9.11). This involves choosing $S_1, S_2, \cdots, S_{B-1}$ in a balanced way, "balance" referring to a symmetry condition between the selected subsets. (Essentially, the set of vectors $\mathbf{1}, \boldsymbol{\delta}_1, \boldsymbol{\delta}_2, \cdots, \boldsymbol{\delta}_{B-1}$, relating to the subsets as in (9.7), has to be such that any one vector has the same set of angles with the remaining $B-1$.) In practice the advantage of balanced subsets over randomly selected ones seems modest, and we will consider only the latter.

Random subsampling is a resampling plan, as described in § 6.1. The choice of a single random subsample S amounts to the choice of a resampling vector $\mathbf{P}^*$ as follows: let

$$I_i^* = \begin{Bmatrix} 1 \\ 0 \end{Bmatrix} \text{prob } \tfrac{1}{2} \quad \text{independently}, \qquad i = 1, 2, \cdots, n.$$

Then, conditional on the event $\sum_{i=1}^n I_i^* > 0$,

$$P_i^* = \frac{I_i^*}{\sum_{j=1}^n I_j^*}. \tag{9.12}$$

9.4. Resampling asymptotics. Random subsampling belongs to a large class of resampling methods, including the bootstrap and half-sampling, which have identical asymptotic properties, at least to a first order of approximation. Consider an arbitrary resampling plan in which we only assume that the components of $\mathbf{P}^*$ are exchangeable. Let $\text{Var}_* P_1^*$ be the variance of any one component under the resampling plan. Notice that

$$0 = \text{Var}_* \sum_{i=1}^n P_i^* = n \text{ Var}_* P_1^* + n(n-1) \text{ Cov}_* (P_1^*, P_2^*).$$

We see that $\mathbf{P}^*$ has mean vector and covariance matrix

$$\mathbf{P}^* \sim \left(\mathbf{P}^\circ, \frac{n}{n-1}(\mathbf{I} - n\mathbf{P}^{\circ\prime}\mathbf{P}^\circ) \text{ Var}_* P_1^*\right), \tag{9.13}$$

$\mathbf{P}^\circ = (1, 1, \cdots, 1)/n$ as before.

Suppose that $\mathscr{X} = \mathscr{R}^1$ and that we are resampling the average $\bar{X}$. Given $\mathbf{X} = \mathbf{x}$, the resampled average, $\bar{X}^* = \mathbf{P}^*\mathbf{x}'$, has mean and variance

$$\bar{X}^* \sim \left(\bar{x}, \frac{n}{n-1} \sum (x_i - \bar{x})^2 \text{ Var}_* P_1^*\right). \tag{9.14}$$

under the resampling distribution.

The bootstrap. $\text{Var}_* P_1^* = (n-1)/n^3$, so $\text{Var}_* \bar{X}^* = \sum (x_i - \bar{x})^2/n^2$.

Random subsampling. It is easy to show, using (9.12), that $\text{Var}_* P_1^* = (n+2)/n^3[1 + o(1/n)]$, so

$$\text{Var}_* \bar{X}^* = \frac{n+2}{n-1} \frac{\sum (x_i - \bar{x})^2}{n^2} \left[1 + o\left(\frac{1}{n}\right)\right]. \tag{9.15}$$

Random half-sampling. Randomly choosing subsamples of size $n/2$, as in line 9 of Table 5.2, gives $\text{Var}_* P_1^* = 1/n^2$ and so $\text{Var}_* \bar{X}^* = \Sigma (x_i - \bar{x})^2/(n \cdot (n-1))$, the usual estimate of variance for a sample average.

The point here is that any resampling plan having $\mathbf{P}^*$ exchangeable and $\lim n^2 \text{ Var}_* P_1^* = 1$ gives, asymptotically, the same value of $\text{Var}_* \bar{X}^*$. This equivalence extends beyond averages to a wide class of smoothly defined random quantities. Working in the finite sample space context of § 5.6, the resampling distribution of $\hat{\mathbf{f}}^*$ is approximately normal,

$$\hat{\mathbf{f}}^* - \hat{\mathbf{f}} \to \mathcal{N}_L\left(0, \frac{\Sigma_{\hat{\mathbf{f}}}}{n-1}(n^2 \operatorname{Var}_* P_1^*)\right)$$

(the expressions for the mean vector and covariance matrix being exact). Smoothly defined random quantities $Q(\hat{\mathbf{f}}^*, \hat{\mathbf{f}})$ have the same limiting distribution under any resampling plan satisfying $\lim_{n\to\infty} n^2 \operatorname{Var}_* P_1^* = 1$.

9.5. Random subsampling for other problems. Line 13 of Table 5.2 shows random subsampling applied to estimate Sd $(\hat{\rho})$ and Sd $(\hat{\phi})$ in the correlation experiment. For each of the 200 trials, $B = 128$ random subsamples were generated as in (9.12), the corresponding values $\hat{\rho}^{*1}, \cdots, \hat{\rho}^{*B}$ computed, and the standard deviation of $\hat{\rho}$ estimated by

$$\widehat{\text{SD}}_{\text{SUB}}(\hat{\rho}) = \left\{ \sum_{b=1}^{B} \frac{[\hat{\rho}^{*b} - \hat{\rho}^{*\cdot}]^2}{[B-1]} \right\}^{1/2} \tag{9.16}$$

(with a similar calculation for $\widehat{\text{SD}}_{\text{SUB}}(\hat{\phi})$).

The results border on the disastrous, especially for $\hat{\phi}$. They would have been even worse if we had not placed a restriction on the subsampling: only subsamples of size $\geqq 4$ were allowed. Asymptotically, we know that $\widehat{\text{SD}}_{\text{SUB}}$ is equivalent to $\widehat{\text{SD}}_{\text{BOOT}}$. Obviously the asymptotics cannot be trusted to predict small sample behavior, at least not in this problem.

Line 14 of Table 5.2 used the same data as line 13, the random subsample values of $\hat{\rho}$ and $\hat{\phi}$, but calculated standard deviations in a more robust way,

$$\widehat{\text{SD}}(\dot{\hat{\rho}}) = \frac{\hat{\rho}^*_{(B_2)} - \hat{\rho}^*_{(B_1)}}{2}, \tag{9.17}$$

$B_1 = [.16(B+1)]$, $B_2 = [.84(B+1)]$, where $\hat{\rho}^*_{(j)}$ is the jth ordered value of $\hat{\rho}^{*1}$, $\hat{\rho}^{*2}, \cdots, \hat{\rho}^{*B}$. In other words, (9.17) is one-half the length of what would be the central 68% confidence interval for ρ, if the typical value theorem applied to this case. (We could just as well apply (9.17) to the bootstrap if we thought that occasional outlying values of $\hat{\rho}^{*i}$ were having an inordinate effect on formula (5.6).) The results are better, but still not encouraging. Another correction, suggested by comparison of (9.15) with the corresponding bootstrap calculation, is to multiply (9.17) by $[(n-1)/(n+2)]^{1/2} = .901$. This gives $\sqrt{\text{MSE}}$ of .083 for estimating Sd $(\hat{\rho})$, and .072 for Sd $(\hat{\phi})$, quite reasonable results, but suspect because of the special "corrections" required.

The problem of choosing among asymptotically equivalent resampling plans is of considerable practical importance. The author feels that the bootstrap has demonstrated some measure of superiority, probably because it is the nonparametric MLE, but the question is still far from settled.

CHAPTER 10

Nonparametric Confidence Intervals

So far we have mainly concentrated on estimating the bias and standard deviation of a point estimator $\hat{\theta}$ for a real parameter θ. This is often all that is needed in applications. However a *confidence interval* for θ is usually preferable. This section, which is highly speculative in content, concerns setting approximate confidence intervals in small sample nonparametric situations.

We begin on familiar ground: setting a confidence interval for the median of a distribution F on the real line. The typical value theorem reduces to the standard order statistic intervals for the median in this case. The bootstrap distribution of the sample median is derived, this being one case where theoretical calculation of the bootstrap distribution is possible. It is shown that the percentiles of the bootstrap distribution also provide (almost) the classical confidence intervals for the median. The method of using the bootstrap distribution, called the *percentile method*, is justified from various theoretical points of view, and improvements suggested. The chapter ends with a brief discussion of more adventurous bootstrap methods for obtaining confidence intervals.

10.1. The median. Let F be a distribution on $\mathscr{R}^1$ with median θ, defined as $\theta = \inf_t [\text{Prob}_F\{X \leqq t\} = .5]$. For convenience we assume that F is continuous. Having observed an i.i.d. sample $X_i = x_i$, $i = 1, 2, \cdots, n$ from F, we can construct exact confidence intervals for θ using the order statistics $x_{(1)} < x_{(2)} < \cdots < x_{(n)}$. Define

$$b_{k,n}(p) = \binom{n}{k} p^k (1-p)^{n-k}, \tag{10.1}$$

the binomial probability of observing k heads in n independent flips of a coin having probability p of heads. The random variable

$$Z = \#\{X_i < \theta\} \tag{10.2}$$

has a binomial distribution with $p = \frac{1}{2}$, $Z \sim \text{Bi}(n, \frac{1}{2})$. Therefore

$$\text{Prob}_F\{x_{(k_1)} < \theta \leqq x_{(k_2)}\} = \sum_{k=k_1}^{k_2-1} b_{k,n}(.5) \tag{10.3}$$

sincce the event $\{x_{(k_1)} < \theta \leqq x_{(k_2)}\}$ is the same as the event $\{k_1 \leqq Z < k_2\}$.

As an example of the use of (10.3), take $n = 13$, $k_1 = 4$, $k_2 = 10$. Then a binomial table gives

$$\text{Prob}_F\{x_{(4)} < \theta \leqq x_{(10)}\} = .908. \tag{10.4}$$

The two tail probabilities are equal, $\text{Prob}_F\{\theta \leqq x_{(4)}\} = \text{Prob}\{Z \leqq 3\} = .046$, and $\text{Prob}_F\{\theta > x_{(10)}\} = \text{Prob}\{Z \geqq 10\} = .046$. In this case $(x_{(4)}, x_{(10)}]$ is a central 90.8% confidence interval for θ.

10.2 Typical value theorem for the median. The median is an m-estimator as described in § 9.1 with $\psi(z) = \text{sgn}(z)$. Since $\psi(z)$ is not strictly monotonic, and we have made no assumptions about the symmetry of F, the typical value theorem of § 9.2 does not apply. However a version of the theorem does hold in this case, as we shall see, and in fact reduces exactly to the binomial interval (10.3).

For any nonempty subset S of $\{1, 2, \cdots, n\}$ let $\hat{\theta}_S$ be the sample median of $\{x_i, i \in S\}$,

$\hat{\theta}_S$ = middle order statistic if $\#S$, the number of elements in S, is odd,

$\hat{\theta}_S$ = any number between the two middle order statistics if $\#S$ is even.

("Between" $x_{(a)}$ and $x_{(b)}$ means lying in the interval $(x_{(a)}, x_{(b)}]$.) Define the random variable

$$Y = \#\{S: \hat{\theta}_S < \theta\}, \tag{10.5}$$

the number of nonempty subsets S for which the sample median is less than the true median. Definition (10.5) assigns Y a *range* of integer values depending on how the even sized cases are assigned.

Example 10.1. Suppose $n = 4$ and $Z = \#\{X_i < \theta\} = 2$, i.e., $\theta \in (x_{(2)}, x_{(3)}]$. Then there are $2^4 - 1 = 15$ nonempty subsets S, of which 5 have $\hat{\theta}_S < \theta$, namely $\{x_{(1)}\}$, $\{x_{(1)}, x_{(2)}\}$, $\{x_{(1)}, x_{(2)}, x_{(3)}\}$, $\{x_{(1)}, x_{(2)}, x_{(4)}\}$ and $\{x_{(2)}\}$. There are 5 sets S for which $\hat{\theta}_S \geqq \theta$, namely $\{x_{(3)}\}$, $\{x_{(3)}, x_{(4)}\}$, $\{x_{(2)}, x_{(3)}, x_{(4)}\}$, $\{x_{(1)}, x_{(3)}, x_{(4)}\}$ and $\{x_{(4)}\}$. Finally, there are 5 ambiguous sets S, namely $\{x_{(1)}, x_{(3)}\}$, $\{x_{(2)}, x_{(3)}\}$, $\{x_{(1)}, x_{(4)}\}$, $\{x_{(2)}, x_{(4)}\}$ and $\{x_{(1)}, x_{(2)}, x_{(3)}, x_{(4)}\}$, for which we can have either $\hat{\theta}_S < \theta$ or $\hat{\theta}_S \geqq \theta$. In this case Y takes on the range of values $\{5, 6, 7, 8, 9, 10\}$, depending on how the ambiguous cases are assigned. In general we have the following relation between the random variables $Z = \#\{X_i < \theta\}$ and $Y = \#\{S: \hat{\theta}_S < \theta\}$:

THEOREM 10.1. *The event $\{Z = z\}$ is equivalent to the event*

$$\sum_{j=0}^{z-1} \binom{n}{j} \leqq Y < \sum_{j=0}^{z} \binom{n}{j}. \tag{10.6}$$

(The proof left to the reader.) In other words, Y is really the same random variable as Z, except that it takes its values in a more complicated space. With $n = 4$, for example, the values $Z = 0, 1, 2, 3, 4$ correspond to $Y = \{0\}, \{1, 2, 3, 4\}, \{5, 6, 7, 8, 9, 10\}, \{11, 12, 13, 14\}, \{15\}$, respectively.

Theorem 10.1 is a version of the typical value theorem. To see this, notice that any postulated value of θ assigns a value to Z, say $z = \#\{x_i < \theta\}$. The attained one-sided significance level of $Z = z$, according to the binomial distribution

$Z \sim \text{Bi}(n, \frac{1}{2})$, satisfies

$$\sum_{j=0}^{z-1} \binom{n}{j} \frac{1}{2^n} < \text{sig level} \leq \sum_{j=0}^{z} \binom{n}{j} \frac{1}{2^n}, \tag{10.7}$$

depending on how the atom at z is assigned.

Going back to the situation described in §§ 9.1 and 9.2, let $Y = \#\{\hat{\theta}_S < \theta\}$. Any postulated value of θ assigns a value y to Y, defined by $\theta \in \mathcal{I}_{y+1}$. The attained one-sided significance level according to the typical value theorem is

$$\frac{y}{2^n} > \text{sig level} \leq \frac{y+1}{2^n}, \tag{10.8}$$

since y of the intervals $\mathcal{I}_j$ lie to the left of θ.

In the case of the median we can't observe $Y = y$ exactly, but rather a range of values y depending on $Z = z$, namely $\sum_{j=0}^{z-1} \binom{n}{j} \leq y < \sum_{j=0}^{z} \binom{n}{j}$. If (10.8) applied here, this range of y values would correspond to the range of significance levels (10.7), which is exactly what we get from the binomial theory. In other words, the typical value theorem does apply to the median, in somewhat coarser form. This coarseness is due to the fact that for the median the $2^n - 1$ choices of S don't correspond to $2^n - 1$ different values of $\hat{\theta}_S$.

It is not surprising that the typical value theorem for the median does not require F to be symmetric about its true median θ. If F is not symmetric about θ to begin with, we can symmetrize it with a monotonic transformation of the data. This transformation doesn't affect the value of Z or Y.

10.3. Bootstrap theory for the median. The bootstrap distribution for the sample median can be calculated theoretically, without recourse to Monte Carlo methods. It is convenient to consider odd sample sizes, say $n = 2m - 1$. Then the sample median $\hat{\theta}$ equals $x_{(m)}$, the middle order statistic.

The bootstrap sample $X_1^*, X_2^*, \cdots, X_n^* \overset{\text{iid}}{\sim} \hat{F}$ has bootstrap sample median $\hat{\theta}^* = X_{(m)}^*$, the mth ordered value of the X_i^*. (Notice that this is true even though there are ties among the bootstrap observations.) Define

$$M_j^* = \#\{X_i^* = x_{(j)}\}, \tag{10.9}$$

$j = 1, 2, \cdots, n$. The event $\{X_{(m)}^* > x_{(k)}\}$ is equivalent to $\{\sum_{j=1}^{k} M_j^* \leq m-1\}$, so that

$$\begin{aligned} \text{Prob}_* \{\hat{\theta}^* > x_{(k)}\} &= \text{Prob}_* \left\{ \sum_{j=1}^{k} M_j^* \leq m-1 \right\} \\ &= \text{Prob}\left\{ \text{Bi}\left(n, \frac{k}{n}\right) \leq m-1 \right\} = \sum_{j=0}^{m-1} b_{j,n}\left(\frac{k}{n}\right). \end{aligned} \tag{10.10}$$

Here we are using $\sum_{j=1}^{k} M_j^* \sim \text{Bi}(n, k/n)$ (see (6.6)) and definition (10.1). Therefore the bootstrap distribution of $\hat{\theta}^*$ is concentrated on the values $x_{(1)} < x_{(2)} < \cdots < x_{(n)}$, with bootstrap probability, say, $p_{(k)}$ of being equal to $x_{(k)}$,

$$p_{(k)} \equiv \text{Prob}_* \{\hat{\theta}^* = x_{(k)}\} = \sum_{j=0}^{m-1} \left\{ b_{j,n}\left(\frac{k-1}{n}\right) - b_{j,n}\left(\frac{k}{n}\right) \right\}. \tag{10.11}$$

Example 10.2. For $n = 13$ the bootstrap distribution is as follows:

$p_{(k)}$	.0000	.0015	.0142	.0550	.1242	.1936	.2230
k	1	2	3	4	5	6	7
$p_{(k)}$	.1936	.1242	.0550	.0142	.0015	.0000	
k	8	9	10	11	12	13	

The bootstrap estimate of standard deviation is $\hat{\sigma}_{\text{BOOT}} = [\sum p_{(k)} x_{(k)}^2 - (\sum p_{(k)} x_{(k)})^2]^{1/2}$. This can be shown to be asymptotically consistent for the true standard deviation of $\hat{\theta}$, in contrast to the jackknife Sd estimate, § 3.4.

10.4. The percentile method. We now discuss a simple method for assigning approximate confidence intervals to any real-valued parameter $\theta = \theta(F)$, based on the bootstrap distribution of $\hat{\theta} = \theta(\hat{F})$. Once we have introduced the method, we will apply it to the case of the median and obtain, almost, the binomial result (10.3).

Let

$$\widehat{\text{CDF}}(t) = \text{Prob}_* \{\hat{\theta}^* \leqq t\} \tag{10.12}$$

be the cumulative distribution function of the bootstrap distribution of $\hat{\theta}^*$. (If the bootstrap distribution is obtained by the Monte Carlo method then $\widehat{\text{CDF}}(t)$ is approximated by $\#\{\hat{\theta}^{*b} \leqq t\}/B$.) For a given α between 0 and .5, define

$$\hat{\theta}_{\text{LOW}}(\alpha) = \widehat{\text{CDF}}^{-1}(\alpha), \qquad \hat{\theta}_{\text{UP}}(\alpha) = \widehat{\text{CDF}}^{-1}(1-\alpha), \tag{10.13}$$

usually denoted simply $\hat{\theta}_{\text{LOW}}$, $\hat{\theta}_{\text{UP}}$. The *percentile method* consists of taking

$$[\hat{\theta}_{\text{LOW}}(\alpha), \hat{\theta}_{\text{UP}}(\alpha)] \tag{10.14}$$

as an approximate $1-2\alpha$ central confidence interval for θ. Since $\alpha = \widehat{\text{CDF}}(\hat{\theta}_{\text{LOW}})$, $1-\alpha = \widehat{\text{CDF}}(\hat{\theta}_{\text{UP}})$, the percentile method interval consists of the central $1-2\alpha$ proportion of the bootstrap distribution.

As an example, consider the law school data, as given in Table 2.1. The bootstrap distribution is displayed in Fig. 5.1. A large number of bootstrap replications, $B = 1{,}000$ in this case, is necessary to get reasonable accuracy in the tails of the distribution. For $\alpha = .16$, the central $1-2\alpha = .68$ percentile confidence interval for ρ is $[.66, .91] = [\hat{\rho} - .12, \hat{\rho} + .13]$. This differs noticeably from the standard normal theory confidence interval for ρ, $[.62, .87] = [\hat{\rho} - .16, \hat{\rho} + .09]$, which is skewed to the left relative to the observed value $\hat{\rho} = .78$. (The normal theory interval has endpoints $\tanh[\hat{\phi} - \hat{\rho}/(2(n-1)) \pm z_\alpha/\sqrt{n-3}]$ where $\hat{\phi} = \tanh^{-1}\hat{\rho}$ and z_α is the $1-\alpha$ point for a standard normal, $z_{.16} = 1$; it is an approximate inversion of the confidence interval for ϕ based on $\hat{\phi} \sim \mathcal{N}(\phi + \rho/(2(n-1)), 1/(n-3))$.) Section 10.7 suggests a bias correction for the percentile method which rectifies this disagreement.

The results of a small Monte Carlo experiment are reported in Table 10.1. 100 trials of $n = 15$ independent bivariate normal observations were generated, true $\rho = .5$. For each trial the bootstrap distribution of $\hat{\rho}^*$ was approximated by $B = 1,000$ bootstrap replications. We see, for example, that in 22 of the 100 trials the true value .5 lay in the region $[\widehat{\mathrm{CDF}}^{-1}(.25), \widehat{\mathrm{CDF}}^{-1}(.5)]$, compared to the expected number 25 if the percentile method were generating exact confidence intervals.

Table 10.1 is reassuring, perhaps misleadingly so when viewed in conjunction with Table 10.2. Central 68% confidence intervals ($\alpha = .16$), with $\hat{\rho}$ subtracted from each endpoint, are presented for the first ten trials of the Monte Carlo experiment. Compared to the normal theory intervals, which are correct here,

TABLE 10.1

100 trials of $X_1, X_2, \cdots, X_{15} \sim$ bivariate normal with $\rho = .5$. For each trial the bootstrap distribution of $\hat{\rho}^*$ was calculated, based on $B = 1,000$ bootstrap replications. In 13 of the 100 trials, the true value .5 lay in the lower 10% of the bootstrap distribution, etc.

Region	0–10%	10–25%	25–50%	50–75%	75–90%	90–100%
Expected #	10	15	25	25	15	10
Observed #	13	16	22	27	12	10

TABLE 10.2

Central 68% confidence intervals for the first ten trials of the Monte Carlo experiment, each interval having $\hat{\rho}$ subtracted from both endpoints.

Trial	$\hat{\rho}$	Normal theory	Percentile method	Bias-corrected percentile method	Smoothed and bias-corrected percentile method
1	.16	(−.29, .26)	(−.29, .24)	(−.28, .25)	(−.28, .24)
2	.75	(−.17, .09)	(−.05, .08)	(−.13, .04)	(−.12, .08)
3	.55	(−.25, .16)	(−.24, .16)	(−.34, .12)	(−.27, .15)
4	.53	(−.26, .17)	(−.16, .16)	(−.19, .13)	(−.21, .16)
5	.73	(−.18, .10)	(−.12, .14)	(−.16, .10)	(−.20, .10)
6	.50	(−.26, .18)	(−.18, .18)	(−.22, .15)	(−.26, .14)
7	.70	(−.20, .11)	(−.17, .12)	(−.21, .10)	(−.18, .11)
8	.30	(−.29, 23)	(−.29, .25)	(−.33, .24)	(−.29, .25)
9	.33	(−.29, .22)	(−.36, .24)	(−.30, .27)	(−.30, .26)
10	.22	(−.29, .24)	(−.50, .34)	(−.48, .36)	(−.38, .34)
Ave	.48	(−.25, .18)	(−.21, .19)	(−.26, .18)	(−.25, .18)

the percentile method gives somewhat erratic results, both in terms of the length of the intervals and of their skewness about $\hat{\rho}$. (Four of the ten percentile intervals are symmetric or even skewed to the right.) The bias corrected percentile method of 10.7 performs better. The final column, which combines smoothing, as in (5.8), with the bias correction is more satisfactory still, but is suspect since the smoothing biases the answers toward what we know is the correct model in this situation.

The percentile method is not as trustworthy as $\hat{\sigma}_{\text{BOOT}}$, which, in the author's experience, can be relied upon to automatically give quite reasonable estimates. On the other hand, setting confidence intervals is a harder problem than estimating standard deviations. The percentile method, perhaps modified as in Table 10.2, is usually more informative than the naive interval $\hat{\theta} \pm c_\alpha \hat{\sigma}$, where c_α is a number taken from the normal or t-tables, although it requires more bootstrap sampling than does the estimation of $\hat{\sigma}$. Some theoretical justification for the percentile intervals is given in §§ 10.5–10.9. More ambitious methods are discussed in § 10.10.

10.5. Percentile method for the median. In the case where $\theta(F)$ is the median of a distribution F on the real line, and $\hat{\theta}$ is the sample median, the percentile method comes very close to giving the classical binomial intervals (10.3). For instance consider the case $n = 13$. The bootstrap distribution of $\hat{\theta}^*$ is supported on the order statistics $x_{(k)}$, as shown at the end of § 10.3, so any percentile interval will be of the form $[x_{(k_1)}, x_{(k_2)}]$. Take $k_1 = 4$ and $k_2 = 10$. The interval $[x_{(4)}, x_{(10)}]$ is a central $1-2\alpha$ interval of the bootstrap distribution with $\alpha = (.0000 + .0015 + .0142 + .0550/2) = .0432$. Here we have split the bootstrap probability at the endpoint of the interval, for reasons discussed at the end of this section. The percentile method assigns

$$[\hat{\theta}_{\text{LOW}}, \hat{\theta}_{\text{UP}}] = [x_{(k_1)}, x_{(k_2)}] \tag{10.15}$$

approximate confidence level $1 - 2\alpha = .914$ for θ. This compares remarkably well with (10.4). Numerical investigation confirms that the agreement is always very good as long as $\alpha \geqq .01$. A theoretical reason for this agreement is given next.

We consider just the lower limit of the interval, the argument for the upper limit being the same. For the classical binomial interval of § 10.1, the α-level connected with the event $\{\theta \leqq x_{(k_1)}\}$ is

$$\alpha = \text{Prob}\,\{Z \leqq k_1 - 1\} = \text{Prob}\,\{\text{Bi}\,(n, \tfrac{1}{2}) \leqq k_1 - 1\}. \tag{10.16}$$

Looking at (10.10), notice that $\text{Prob}_*\,\{\hat{\theta}^* \leqq x_{(k_1)}\} = \text{Prob}\,\{\text{Bi}\,(n, k_1/n) \geqq m\}$. The percentile method for the median, taking into account splitting the bootstrap probability at the endpoint, amounts to assigning significance level $\hat{\alpha}$ to $\{\theta \leqq x_{(k_1)}\}$ where

$$\hat{\alpha} = \frac{1}{2}\left[\text{Prob}\left\{\text{Bi}\left(n, \frac{k_1}{n}\right) \geqq m\right\} + \text{Prob}\left\{\text{Bi}\left(n, \frac{k_1 - 1}{n}\right) \geqq m\right\}\right]. \tag{10.17}$$

If the reader replaces (10.16) and (10.17) with their usual normal approximations, making the continuity corrections, but ignoring the difference in denominators, he will see why $\hat{\alpha}$ approximates α. (Actually the approximation is mysteriously better than this computation suggests, especially when k_1/n is much less than $\frac{1}{2}$.)

Let

$$D(t) = \text{Prob}_F\,\{X < t\}, \qquad \hat{D}(t) = \frac{\#\{x_i < t\}}{n} \tag{10.18}$$

be the cumulative and empirical cumulative distribution functions. Then

$$\hat{D}(t) \sim \frac{\text{Bi}(n, D(t))}{n}, \tag{10.19}$$

so if $D(t)=\frac{1}{2}$ then $\hat{D}(t) \sim \text{Bi}(n, \frac{1}{2})/n$. According to (10.16), another way to describe the binomial α-level interval is the following: the lower limit of the interval is the smallest value of t for which we can accept the null hypothesis $D(t)=\frac{1}{2}$ with one-sided significance level α. The lower limit of the percentile interval has this interpretation: it is the smallest value of t for which $\text{Prob}\{\text{Bi}(n, \hat{D}(t))/n > \frac{1}{2}\} > \alpha$. In other words, the binomial interval checks whether $\hat{D}(t)$ is too small compared to expectation $\frac{1}{2}$, while the percentile interval checks whether $\frac{1}{2}$ is too large compared to expectation $\hat{D}(t)$.

It is not surprising that the percentile method for the median requires splitting the bootstrap probability at the endpoints. The problem is the same as in the typical value theory for the median, namely that the sample median takes on only n possible different values under bootstrap sampling. This contrasts with smoothly defined statistics, such as the correlation, for which the bootstrap distribution is effectively continuous when $n \geqq 10$. Suppose that instead of the sample median we were considering the m-estimator $\hat{\theta}_c$ of Example 9.2 with c very large. Then if a bootstrap median $\hat{\theta}^*$ equals $x_{(k)}$, the corresponding value of $\hat{\theta}_c^*$ will be *almost but not quite* equal to $x_{(k)}$. Take $n=13$ and $\alpha=.0432$ as in (10.15). Then $\lim_{c\to\infty} \hat{\theta}_{\text{LOW},c} = x_{(4)}$, and $\lim_{c\to\infty} \hat{\theta}_{\text{UP},c} = x_{(10)}$. In this sense, $[x_{(4)}, x_{(10)}]$ is a .914 confidence interval for the median, as claimed. The Bayesian arguments of the next section provide another justification for splitting the endpoint probabilities.

10.6. Bayesian justification of the percentile method. We assume that the sample space $\mathscr{X}$ is discrete as in § 5.6. This is no real restriction, since we can take the number L of discrete categories arbitrarily large. If $\mathscr{X}$ is the real line, for instance, we might partition $[-10^{10}, 10^{10}]$ into $2\cdot 10^{20}$ intervals of length 10^{-10} Then $L = 2\cdot 10^{20}+2$, counting the semi-infinite end intervals, and for practical purposes the discretization will have no effect on our inferences. As in § 5.6, we let f_l equal the probability that X occurs in category l, with $\hat{f}_l$ equal to the corresponding observed frequency $\#\{x_i \in \text{category } l\}/n$, and denote $\mathbf{f}=(f_1, f_2, \cdots, f_L)$, $\hat{\mathbf{f}}=(\hat{f}_1, \hat{f}_2, \cdots, \hat{f}_L)$.

We take the prior distribution on $\mathbf{f}$ to be a symmetric Dirichlet distribution with parameter a,

$$\mathbf{f} \sim \text{Di}_L(a\mathbf{1}), \tag{10.20}$$

i.e., the prior density of $\mathbf{f}$ is taken proportional to $\prod_l f_l^{a-1}$. Having observed $\hat{\mathbf{f}}$, the a posteriori density of $\mathbf{f}$ is

$$\mathbf{f}|\hat{\mathbf{f}} \sim \text{Di}_L(a\mathbf{1}+n\hat{\mathbf{f}}),$$

with density function proportional to $\prod_l f_l^{n\hat{f}_l+a-1}$. Letting $a\to 0$ to represent prior ignorance gives the well-known result

(10.21) $$\mathbf{f}|\hat{\mathbf{f}} \sim \mathrm{Di}_L\,(n\hat{\mathbf{f}}).$$

Distribution (10.21) is quite similar to the bootstrap distribution (5.13),

(10.22) $$\mathbf{f}^*|\hat{\mathbf{f}} \sim \frac{\mathrm{Multi}_L\,(n, \hat{\mathbf{f}})}{n}.$$

(i) Both distributions are supported entirely on those categories having $\hat{f}_l > 0$, i.e., on those categories in which data were observed.

(ii) Both distributions have expectation vector $\hat{\mathbf{f}}$.

(iii) The covariance matrices are also nearly equal, $\mathrm{Cov}\,(\mathbf{f}|\hat{\mathbf{f}}) = \Sigma_{\hat{f}}/(n+1)$, $\mathrm{Cov}_*\,(\hat{\mathbf{f}}^*|\hat{\mathbf{f}}) = \Sigma_{\hat{f}}/n$, where $\Sigma_{\hat{f}}$ has diagonal elements $\hat{f}_l(1-\hat{f}_l)$ and off-diagonal elements $-\hat{f}_l\hat{f}_m$.

The point here is that the a posteriori distribution of $\theta(\mathbf{f})|\hat{\mathbf{f}}$ is likely to be well approximated by the bootstrap distribution of $\theta(\hat{\mathbf{f}}^*)|\hat{\mathbf{f}}$, if $\theta(\mathbf{f})$ is any reasonably smooth function of $\mathbf{f}$. If this is true, the percentile method $1-2\alpha$ central confidence interval will be a good approximation to the central Bayes interval of probability $1-2\alpha$.

The prior distribution $\mathrm{Di}_L\,(\alpha\mathbf{1})$, $\alpha \to 0$, may seen unreasonable,[1] but it gives a reasonable answer when $\theta(\mathbf{f})$ is the median of a distribution on the real line. In this case, letting the discretization of $\mathscr{X}$ become infinitely fine, it can be shown that

(10.23) $$\mathrm{Prob}\,\{x_{(k_1)} < \theta(\mathbf{f}) \leqq x_{(k_2)}|\hat{\mathbf{f}}\} = \sum_{k=k_1}^{k_2-1} b_{k,n-1}(.5),$$

comparing nicely with the classic binomial interval (10.3).

Realistically we would never believe that our a posteriori distribution for $\mathbf{f}$ concentrates exclusively on just the data points already seen. Smoothing the distribution (10.21) even slightly splits the endpoint probabilities in (10.23), as in the application of the percentile method to the median.

10.7. The bias-corrected percentile method. The bootstrap distribution for the sample median, § 10.3, is *median unbiased* in the sense that $\mathrm{Prob}_*\,\{\hat{\theta}^* \leqq \theta\} = .50$ (splitting the probability at $\hat{\theta} = x_{(m)}$). The argument which follows suggests that if $\mathrm{Prob}_*\,\{\hat{\theta}^* \leqq \hat{\theta}\} \neq .50$ then a bias correction to the percentile method is called for.

To be specific, define

(10.24) $$z_0 = \Phi^{-1}(\widehat{\mathrm{CDF}}\,(\hat{\theta})),$$

where $\mathrm{CDF}\,(t) = \mathrm{Prob}_*\,\{\hat{\theta}^* \leqq t\}$ as in (10.12), and Φ is the cumulative distribution function for a standard normal variate. The *bias corrected percentile method* consists of taking

(10.25) $$[\widehat{\mathrm{CDF}}^{-1}\,(\Phi(2z_0 - z_\alpha)), \widehat{\mathrm{CDF}}^{-1}\,(\Phi(2z_0 + z_\alpha))]$$

[1] In a recent paper (Rubin (1979)) this criticism is made, with the suggestion that it would be better to do the Bayesian analysis using a more informative prior distribution.

as an approximate $1-2\alpha$ central confidence interval for θ. Here z_α is the upper α point for a standard normal $\Phi(z_\alpha)=1-\alpha$.

Notice that if $\text{Prob}_*\{\hat\theta^* \leqq \theta\}=.50$ then $z_0=0$ and (10.25) reduces to (10.14), the uncorrected percentile interval. However, even small differences of $\text{Prob}_*\{\hat\theta^* \leqq \hat\theta\}$ from .50 can make (10.25) much different from (10.14). In the law school data, for example, $\widehat{\text{CDF}}(\hat\rho)=.433$ (i.e., in 433 out of 1,000 bootstrap replications, $\hat\rho^*$ was less than $\hat\rho=.776$). Therefore $z_0=\Phi^{-1}(.433)=-.17$, and taking $\alpha=.16$, $z_\alpha=1$, in (10.25) gives the approximate 68% interval

$$[\widehat{\text{CDF}}^{-1}(\Phi(-1.34)), \widehat{\text{CDF}}^{-1}(\Phi(.66))]=[\widehat{\text{CDF}}^{-1}(.090), \widehat{\text{CDF}}^{-1}(.745)]$$
$$=[\hat\rho-.17, \hat\rho+.10]$$

for ρ. This compares with the uncorrected percentile interval $[\widehat{\text{CDF}}^{-1}(.16), \widehat{\text{CDF}}^{-1}(.84)]=[\hat\rho-.12, \hat\rho+.13]$ and the normal theory interval $[\hat\rho-.16, \hat\rho+.09]$.

The argument supporting (10.25) is based on hypothesizing a transformation to a normal pivotal quantity. Suppose there exists some monotonic increasing function $g(\cdot)$ such that the transformed quantities

$$\phi=g(\theta), \quad \hat\phi=g(\hat\theta), \quad \hat\phi^*=g(\hat\theta^*) \tag{10.26}$$

satisfy

$$\hat\phi-\phi\sim\mathcal{N}(-z_0\sigma, \sigma^2), \qquad \hat\phi^*-\hat\phi \underset{*}{\sim}\mathcal{N}(-z_0, \sigma, \sigma^2) \tag{10.27}$$

for some constants z_0 and σ. In other words, $\hat\phi-\phi$ is a *normal pivotal quantity*, having the same normal distribution under F and $\hat F$. (Remember, the distribution of $R=\hat\phi-\phi$ under $\hat F$ is what we call the bootstrap distribution of $R^*=\hat\phi^*-\hat\phi$, "$\underset{*}{\sim}$" indicating the distribution under i.i.d. sampling from $\hat F$.)

In parametric contexts, (10.27) is a device frequently used to obtain confidence intervals. Fisher's transformation $\phi=\tanh^{-1}\rho$ for the correlation coefficient is the classical example. Within the class of bivariate normal distributions it produces a good approximation to (10.27), with $\sigma^2=1/(n-3)$ and $z_0=-\rho\sqrt{n-3}/(2(n-1))$. The distributions are not perfectly normal, and z_0 is not perfectly constant, but the theory still produces useful intervals, as described in § 10.4.

The middle statement in (10.26) is actually a definition of the estimator $\hat\phi$. If $\hat\theta=\theta(\hat F)$ is a functional statistic, then $\hat\phi=g(\theta(\hat F))$ is the nonparametric maximum likelihood estimator for ϕ. The last relationship in (10.26) follows from $\hat\phi^*=\hat\phi(X_1^*, X_2^*, \cdots, X_n^*)=g(\hat\theta(X_1^*, X_2^*, \cdots, X_n^*))=g(\hat\theta^*)$. It implies that the bootstrap distribution of $\hat\phi^*$ is the obvious mapping of the bootstrap distribution of $\hat\theta^*$. Letting

$$\widehat{\text{CDG}}(s)=\text{Prob}_*\{\hat\phi^*\leqq s\},$$

we have

$$\widehat{\text{CDG}}(g(t))=\widehat{\text{CDF}}(t) \tag{10.28}$$

for all t.

The standard $1-2\alpha$ confidence interval for ϕ is

$$\phi \in [\hat{\phi} + z_0\sigma - z_\alpha\sigma, \hat{\phi} + z_0\sigma + z_\alpha\sigma]. \tag{10.29}$$

Using (10.27), we will see that mapping (10.29) back to the θ scale gives (10.25). First notice that (10.27) and (10.28) imply

$$\text{Prob}_* \{\hat{\phi}^* \leqq \phi\} = \Phi(z_0) = \widehat{\text{CDG}}\,(g(\hat{\theta})) = \widehat{\text{CDF}}\,(\hat{\theta}),$$

which gives $z_0 = \Phi^{-1}(\widehat{\text{CDF}}\,(\hat{\theta}))$ as in (10.24).

Using (10.27) again,

$$\text{Prob}_* \{\hat{\phi}^* + z_0\sigma \pm z_\alpha\sigma\} = \text{Prob}_F \{\hat{\phi} < \phi + z_0\sigma \pm z_\alpha\sigma\} = \Phi(2z_0 \pm z_\alpha),$$

or, since this can be written as $\widehat{\text{CDG}}\,(\hat{\phi} + z_0\sigma \pm z_\alpha\sigma) = \Phi(2z_0 \pm z_\alpha)$,

$$\hat{\phi} + z_0\sigma \pm z_\alpha\sigma = \widehat{\text{CDG}}^{-1}\,[\Phi(2z_0 \pm z_\alpha)].$$

Transforming (10.29) back to the θ scale by the inverse mapping $g^{-1}(\cdot)$ gives the interval with endpoints $g^{-1}(\hat{\phi} + z_0\sigma \pm z_\alpha\sigma) = g^{-1}\,\widehat{\text{CDG}}^{-1}\,[\Phi(2z_0 \pm z_\alpha)] = \widehat{\text{CDF}}^{-1}\,[\Phi(2z_0 \pm z_\alpha)]$ (the last equality following from (10.28), $[\widehat{\text{CDF}}]^{-1} = [\widehat{\text{CDG}}\,g]^{-1} = g^{-1}\,\widehat{\text{CDG}}^{-1}$). We have now derived (10.25) from (10.27).

The normal distribution plays no special role in this argument. Instead of (10.27) we could assume that the pivotal quantity has some other symmetric distribution than normal, in which case Φ would have a different meaning in (10.25). In the unbiased case, where $z_0 = 0$, the normal distribution plays no role at all since we get the uncorrected percentile interval (10.14). This is worth stating separately: *If we assume there exists a monotonic mapping $g(\cdot)$ such that $\hat{\phi} - \phi$ and $\hat{\phi}^* - \hat{\phi}$ have the same distribution, symmetric about the origin, then the percentile interval* (10.14) *has the correct coverage probability.*

None of these arguments require knowing the form of the transformation g, only of its existence. Consider the correlation coefficient again, assuming that the true distribution F is bivariate normal. Applying the parametric bootstrap of § 5.2, and following definition (10.25), will automatically give almost exactly the normal theory interval $\tanh[\hat{\phi} - \hat{\rho}/(2(n-1)) \pm z_\alpha/\sqrt{n-3}]$, *without any knowledge of the transformation $\hat{\phi} = \tanh^{-1}\hat{\rho}$.*

10.8. Typical value theory and the percentile method. The uncorrected percentile method (10.13), (10.14) is a direct analogue of typical value theory, as described in § 9.2. In fact, if we let $\widehat{\text{CDF}}\,(t)$ be the cumulative distribution function of the subsample values rather than of the bootstrap values, $\widehat{\text{CDF}}\,(t) = \#\{\hat{\theta}_S \leqq t\}/(2^n - 1)$, then (10.13), (10.14) gives the $1-2\alpha$ central subsample interval. The same connection holds for the Monte Carlo versions of the two methods, where $\widehat{\text{CDF}}$, either subsample or bootstrap defined, is approximated by Monte Carlo simulations.

The asymptotic considerations of § 9.4 suggest that the typical value intervals will be wider than the percentile intervals by a factor of about $\sqrt{(n+2)/(n-1)}$. We see this effect in Table 10.3. Ten i.i.d. samples $X_1, X_2, \cdots, X_{15}$ were obtained from the negative exponential distribution, $\text{Prob}\{X > x\} = e^{-x}$, $x > 0$. Four